SpringerBriefs in Complexity

SpringerBriefs in Complexity are a series of slim high-quality publications encompassing the entire spectrum of complex systems science and technology. Featuring compact volumes of 50 to 125 pages (approximately 20,000–45,000), Briefs are shorter than a conventional book but longer than a journal article. Thus Briefs serve as timely, concise tools for students, researchers, and professionals.

Typical texts for publication might include:

- A snapshot review of the current state of a hot or emerging field
- A concise introduction to core concepts that students must understand in order to make independent contributions
- An extended research report giving more details and discussion than is possible in a conventional journal article,
- A manual describing underlying principles and best practices for an experimental or computational technique
- An essay exploring new ideas broader topics such as science and society

Briefs allow authors to present their ideas and readers to absorb them with minimal time investment. Briefs are published as part of Springer's eBook collection, with millions of users worldwide. In addition, Briefs are available, just like books, for individual print and electronic purchase. Briefs are characterized by fast, global electronic dissemination, straightforward publishing agreements, easy-to-use manuscript preparation and formatting guidelines, and expedited production schedules. We aim for publication 8–12 weeks after acceptance.

SpringerBriefs in Complexity are an integral part of the Springer Complexity publishing program. Proposals should be sent to the responsible Springer editors or to a member of the Springer Complexity editorial and program advisory board (springer.com/complexity).

More information about this series at http://www.springer.com/series/8907

Massimo Lapucci · Ciro Cattuto
Editors

Data Science for Social Good

Philanthropy and Social Impact in a Complex World

 Springer

Editors
Massimo Lapucci
Turin, Italy

Ciro Cattuto
Turin, Italy

ISSN 2191-5326   ISSN 2191-5334   (electronic)
SpringerBriefs in Complexity
ISBN 978-3-030-78984-8   ISBN 978-3-030-78985-5   (eBook)
https://doi.org/10.1007/978-3-030-78985-5

This Springer imprint is published by the registered company Springer Nature Switzerland AG
The registered company address is: Gewerbestrasse 11, 6330 Cham, Switzerland

Preface

The digital transformation promises novel opportunities for positive social change, while also foreshadowing unprecedented challenges and risks. Far from being a technical process driven by market forces alone, the digital transformation is a historical process that touches all aspects of our lives and calls for a deep reflection on the role that our institutions can and should play to ensure that the best possible outcomes are achieved for all.

Data, and the Science of Data, lie at the core of this unfolding transformation: Data affords new ways of measuring, understanding and predicting our world, and is fueling a new wave of technologies and systems—based on Data Science and Artificial Intelligence—for taking better decisions and designing better policies. These capabilities are new and potentially disruptive, hence they need to be closely monitored and governed in the impact they generate on our society. Indeed, none of the technologies enabled by Data are born neutral: power relations, prejudices, inequalities and discrimination are pervasive throughout the value chain of data-driven digital products and systems. The need to govern the digital transformation—so that it produces the best outcomes for citizens—has therefore emerged as one of the central challenges of the current decade. Governments, public agencies, non-governmental organizations, industries and nonprofits are all grappling with the opportunities of using data and data-driven capabilities for their respective missions, as well as with the challenges of enabling the new cross-sectoral collaborations that are often required to unlock the sought value.

This book aims to contribute to the discussion on Data Science and its potential for positive social impact by gathering the perspectives of pioneers and thought leaders in this fast-growing space. Chapter contributors comprise leading philanthropies, think tanks, data initiatives within non-governmental organizations. Through the reflections and experiences of leaders at these organizations, the book showcases success stories on using data for social good, recurring challenges, organizational innovations and future directions, including systemic changes and governance innovations that will be needed to advance the maturity of the data ecosystem and foster the emergence of new kinds of common goods.

Data can impact decisions and policies only when the knowledge it yields is action-able—and acted upon. The readers this book caters to are therefore as diverse as the skills needed to turn data into knowledge, and to then turn knowledge into better deci-sions and policies. Gleaning knowledge from data requires scientific competencies in data science, artificial intelligence, complex systems, as well as domain-specific knowledge from, e.g., the social sciences, public health, demography, etc. In turn, acting on new knowledge to design better decisions and policies requires competen-cies in governance, ethics, finance, law, policy-making, impact evaluation and more. The book, therefore, targets both readers who are researchers, analysts and engi-neers, tasked with making sense of data and designing new technical solutions, and readers who are decision-makers at social impact organizations tasked with adopting data-driven technologies to better execute on their mission. It aims to provide value for both audiences by sharing end-to-end stories of data science for social impact, as well as by helping create mutual awareness and a shared language across these traditionally separated cultures.

Using data for better social impact also critically depends on creating and sustaining cross-sector collaborations between problem owners (social impact orga-nizations, non-governmental organizations, foundations and philanthropies), data owners (corporations, public sector, etc.) and knowledge-intensive organizations (academia, research centers, think tanks). We dare hope that the content of the book will be valuable and inspiring for readers at such diverse organizations. Leaders and program managers at social impact organizations will find success stories and reflections on where we stand and what are the challenges ahead in generating social impact from data. Decision-makers at data-intensive companies will find accounts on how privately-held data have been used to generate new common goods that go well beyond corporate social responsibility. Researchers and students—especially young data scientists—will find inspiration on what their technical skills can achieve if empowered by a social mission, possibly shaping the interests and priorities of their work. Finally, communicators might be stimulated to delve deeper into this emerging space and contribute to the public discourse on data science and its potential for the public interest.

We are honored by the thought leaders who generously agreed to contribute to this project, and we hope that readers will find these chapters as inspiring and thought-provoking as we did.

Turin, Italy   Ciro Cattuto
                                                               Massimo Lapucci

Acknowledgments

An edited book builds on the generosity of chapter contributors and of their organizations. We express our gratitude to Alberto Alemanno at HEC Paris and The Good Lobby, Paula Hidalgo-Sanchis at UN Global Pulse, Stefan Germann and Ursula Jasper at Fondation Botnar, Claudia Juech at the patrick J. McGovern Foundation, Nuria Oliver at DataPop Alliance and the European Laboratory for Intelligent Systems (ELLIS), Rachel Rank at 360Giving, and Stefaan Verhulst at The Governance Laboratory of New York University (GovLab). We thank Stefaan Verhulst for stimulating conversations, introductions and suggestions that helped us design this book. We are also grateful to the several organizations that have enabled our exploration of this space by means of conferences, workshops, events and research projects: CRT Foundation, the European Foundation Centre, ISI Foundation and OGR Torino. Finally, we thank Hisako Niko and Christopher Coughlin at Springer Nature for embracing this project and for their support and guidance from inception to publication.

Contents

About the Editors

Dr. Massimo Lapucci is the Managing Director (Secretary General) of CRT Foundation, a philanthropic foundation based in Italy with an endowment of about three billion euro. He is the Secretary General of "Development and Growth Foundation", a CRT subsidiary focused primarily on impact investing, tech and innovation, and he is the Chief Executive Officer of OGR, a former large industrial space in Turin, Italy, recently converted into an innovative and experimental center for contemporary culture, art, research, and start-ups in partnership with the US accelerators. He was the former Chair of the European Foundation Centre in Bruxelles, the network of institutional philanthropy which unites over 300 members from nearly 40 countries, including the USA. Dr. Lapucci has extensive international experience as an Investment and Finance Officer and a board member for public and private companies in various sectors and non-profit organizations in the EU and the Americas, including the Rockefeller P.A., Europe, the advisory board of the London School of Economics-Marshall Institute; ISI Foundation in Turin, Italy; and ISI Global Science Foundation in New York, NY, USA. He is Vice President of the Social Impact Agenda for Italy and a member of the Global Social Impact Investment Steering Group. Dr. Lapucci also has a consolidated teaching experience at the university level, and since 2006 he is a World Fellow at Yale University, USA

Prof. Ciro Cattuto, PhD is an Associate Professor in the Computer Science Department of the University of Turin, Italy, and a Principal Researcher and Research Area Coordinator at ISI Foundation in Turin, Italy. His interests include Data Science, Complex Systems, Public Health, and the social impact of data. He holds a PhD in Physics from the University of Perugia, Italy, and has carried out interdisciplinary research at the University of Michigan, USA, at the Enrico Fermi Center and Sapienza University in Rome, and at the Frontier Research System of RIKEN, Japan. He is a founder and principal investigator of the SocioPatterns collaboration, an international effort on studying social networks with wearable sensors, with applications to epidemiology. He is an editorial board member of Scientific Data, EPJ Data Science, PeerJ Computer Science, Journal of Computational Social Science, Data and Policy. He was an organizer and chair of leading conferences in Computer Science, Data Science, Network Science, and Complex Systems. He is a Fellow of the European

Laboratory for Learning and Intelligent Systems (ELLIS). He was a member of the COVID-19 task force of the Italian Ministry for Technological Innovation and Digital Transition, and he is a member of the Italian Ministry of Labour's Working Group on Algorithmic Governance.

Introduction

Massimo Lapucci

The momentum behind the topic of Data within the philanthropic sector, and more specifically Data for Good, has been building for some time now. A topic once perhaps left to the margins, it has now taken a more prominent position in the philanthropic landscape.

When I arrived at Yale University as a World Fellow, I remember that I began to explore more how the role of economics could evolve to reconcile a fundamental and fair return to shareholders and investors with a positive social impact. These reflections soon led me to discover a world that would not only shape my future professional choices as an economist and financial manager, but those of the entire world order for years to come.

We live in challenging times. From climate change to economic inequality and forced migration, the difficulties confronting decision makers are unprecedented in their variety, complexity and urgency. If those emergencies were not enough, we have also been forced in 2020 to face the crisis posed by Covid-19, which has obliged us to find new ways of living: from the way in which we work to our social interactions, it no longer seems possible to return to the way we were before.

When faced with such relevant, urgent and universal issues, the philanthropic sector is also responsible for identifying solutions based on scientific approaches and innovative methodologies of analysis, using Data and its science to obtain the analysis and forecasting tools to address some of the major challenges of our society, as encompassed in the UN Agenda 2030. In order to extract the social value from Data, we all, with a common effort, must have clear in our minds the seventeenth Sustainable Development Goal, that of Partnerships. The time is now to find global answers to these global challenges.

The growing interconnection of the world, its measurability through Data, the adoption of increasingly intelligent automatic systems, the profusion of the Internet

M. Lapucci (✉)
Secretary General, Fondazione CRT, Turin, Italy
e-mail: massimo@lapucci.eu

M. Lapucci and C. Cattuto (eds.), *Data Science for Social Good*,
SpringerBriefs in Complexity, https://doi.org/10.1007/978-3-030-78985-5_1

1

of Things, once accumulated make this science highly relevant for everyone, in particular to the philanthropic sector.

As the sector familiarizes itself with these new opportunities, inspiration and also learning can be found within the profit sector, where Data Science is considered transformational in the way it helps large companies make decisions, and instrumental in measuring their success.

It is not just a matter of one-way learning, but rather an opportunity for a reciprocal exchange that paves the way for greater hybridization between sectors, thanks to concerted efforts and targeted actions by several actors, amongst whom Fondazione CRT [1] where I have been serving as Secretary General since 2012, and the European Foundation Centre [2] in Brussels of which I was recently Chair.

As philanthropic institutions, we are called not only to take up but to lead the improvement in the use of Data for social and impact purposes, networking in a national and international dimension, listening to sectors outside our own such as the profit, finance, and environmental sectors towards creating a more sustainable society for all.

The non-profit sector, without distorting itself, has the opportunity to draw inspiration from profit in terms of more efficient use of resources, accountability and transparency, using innovative scientific approaches of Big Data and Artificial Intelligence to identify needs, target interventions and assess impact.

At the same time, the for-profit sector is able to learn how to build its action around the fundamental principles of social and environmental sustainability: a sort of neo-humanism capable, in perspective, of "doing good" to the economy, not only with a view of expanding their activities, but also at the level of reputation, as businesses start to be perceived as a system attentive to improving the quality of life of communities and of people.

A shift in approach towards greater hybridization is already taking place, where more and more profit organizations are taking a stronger position on social and political issues [3].

This hybridization process is a fundamental step towards the development of a more dynamic society, better prepared to face changes and to respond to new needs, reducing the "disconnection" between citizens and public and private institutions.

Even though the for-profit sector has been using big Data and AI for a long time, it is largely shared that the market alone is unable to extract that "social value" through big Data and Data science, of which there is a great need in order to tackle societal challenges.

The European Union has come forward to help bridge this gap.

Firstly, the European Commission has recognized the use of Data for public interest purposes as an integral part of the Strategy for the Digital Single Market, and has launched actions to encourage its dissemination. In addition, there is the implementation of the General Regulation on Data Protection as well as the creation of a High-Level Expert Group on Business-to-Government (B2G) Data sharing. In the current climate, the European Commission has identified the digital transition as one of the three priorities for the post-Covid recovery phase, considering the creation of a Data Economy as an engine for innovation and employment creation.

Thanks to these actions undertaken by European Institutions, Europe has positioned itself as a particularly adept space in which to develop the use of Data for social good, offering at the same time to the philanthropic sector an opportunity to play an important role in this process, spearheading a new agenda for the sector, in alignment with European values.

Amongst the opportunities that Data offers the philanthropic sector is the ability to provide that scientific methodology to our work, that key of knowledge that for philanthropy becomes fundamental to be able to better reach its own mission, which ultimately lies in the public good. Some claim not only a scientific approach to philanthropy, but a real "science for philanthropy" to provide that strong evidence of the effectiveness of donors, which is currently still missing [4].

This takes on greater meaning when we consider the capacity of philanthropy, and how it should perhaps be considered as its own sector to all effects. We need only look to Europe, where we can count upon 147,000 foundations, that provide € 60 billion a year and manage assets of € 511 billion [5], or to the USA where the total giving by foundations was more than $79 billion in 2019 [6], to fully appreciate the robust structure of philanthropy.

Another potential benefit of Data being applied to philanthropy arises from the awareness that human decisions are not perfect. With the use of algorithms in particular, Data can help to better inform decisions, so to improve efficiency, as Nuria Oliver underscores in her contribution.

Today, I believe that for non-profit organizations to consolidate their support, funding decisions can no longer be based solely and exclusively on the merits of the cause: other crucial factors now come into greater consideration, such as their ability to demonstrate measurable and transparent results, as well as a capacity for higher levels of sustainability and efficiency.

This collection of articles from international thought leaders and experts will explore how Data presents a multitude of possibilities and opportunities for the philanthropic sector.

In the following chapters, we will deepen our understanding of how Data contributes to improving situational awareness and response, permitting a greater understanding of linkages between otherwise unassociated phenomena. Data can help in improving predictive and forecasting abilities, as well as monitoring and evaluating the impacts of policies and interventions [7].

The 360 Giving Initiative, founded by UK philanthropist Fran Perrin, is also an excellent example of how through Data sharing the gap of information can be bridged, through the provision of support alongside a technical framework—a Data Standard—with its benefits reaped by both beneficiaries and funding organizations. A bridge that leads to learning, greater digital literacy, knowledge sharing, and ultimately to make better-informed decisions that can improve our overall impact as a sector.

Data can have a significant influence in philanthropic work; it can assist in enriching and elevating impact evidence as part of funding requests, and it can similarly be of service to donors who are increasingly called upon to show a higher level of accountability and prove the specific impact of their investments with measurable and transparent results.

More than ever, philanthropy needs to adopt a series of tools integrated together in a real toolbox capable of acting in a temporally, or functionally, coordinated manner: from the more traditional grant making, to the innovative developments that are emerging in philanthropy. Among these tools are Data science, as well as impact investing—which integrates the typical binomial of traditional risk-return investments, with a third fundamental element: that of Social Impact, understood as the creation of value for society—whose successful use could be further guaranteed by Data. One instrument should not be seen as a substitute or more efficient than the other, but it is in their integration that the maximization of the social value created should be identified, in accordance with the pursued objective. There is no magic formula or philanthropy 2.0 that can replace the previous one, but there are now different and new tools and responses for a rapidly evolving environment.

In other words, Data, when used responsibly and systematically, can become a real differentiator in program effectiveness and impact for decision makers, and among these, philanthropic organizations.

Opportunities aside, Data poses new types of risks and challenges. As we strive to unlock and access the value from Big Data, we must take the necessary precautions to construct a safe framework in which we can operate, taking into consideration the ethical, legal, and regulatory questions posed, with an eye to the future and the questions that will undoubtedly emerge as we move forwards.

Global Pulse, as you can read in these pages, has contributed greatly to this with the creation of a set of practical instruments—including the creation of a Data Privacy Advisory Group—to address the ethical, security, and privacy challenges that we must affront when using big Data for public good.

Because of its mission, that ultimately consist in pursuing the public interest, and because its autonomy, due the financial independence, and freedom from the typical checks and balances of politics and the market, philanthropy, and specifically institutional philanthropy, is particularly well positioned to play a unique and vital role in helping to ensure that the positive potential of Data is unleashed while limiting its possible negative consequences.

While ensuring the public good, I am persuaded that philanthropy not only has a role to play in mitigating the risk associated with the use of Data for public good—for example by stimulating the determination of the contours of the Data market–, but it can also be a key player as a convener, in enabling a greater discourse between the supply (Data holders) and demand (Data users), thus encouraging the collaboration between different actors, associations, NGOs, research centres, businesses, and the private sector as a whole. As Claudia Juech rightly mentions in her chapter, each of us, from citizens, to academics and governments, all have a role to play in the advancement of the field of Data for Good.

A greater collaboration between Data users and Data holders, as well as across different sectors and domains of expertise, increases the real-world positive impact of Data. It also increases trust and ethics in the way Data is handled and, importantly, in the perceived legitimacy of such efforts.

Data collaboration is therefore crucial to unlocking the potential of Data and Data science for good, while at the same time limiting its potential risk and harm.

Collaboration is an area where The GovLab in New York has worked extensively, and notably during the Covid-19 pandemic they launched an international call—also promoted by Fondazione CRT and ISI Foundation—encouraging action aimed at increasing the availability and use of Data to tackle the current global emergency, with an eye on making responses to future societal and environmental threats more efficient, quicker and more prepared. It is encouraging that this call for action reached almost 500 signatures by the end of July 2020. If I had to identify a positive effect from the recent pandemic, I think I can say that it is a greater availability in Data sharing, as shown by some individual collaboration projects. The challenge now is to make these collaborations systemic and structured, in a framework of trust and transparency. And the Covid emergency has now placed us before a fundamental topic for our future: the need to immediately share knowledge bases, experiences, and therefore the information Data on the phenomena observed and detected. This is not science fiction or even political fantasy, but a necessary and possible step that the whole world has to take. A leap forward in dealing not only with emergencies, but also more generally with globalization and its challenges.

In their chapter, Stefan Germann and Ursula Jasper thoroughly explore the application of Data science, digital technology and artificial intelligence to healthcare systems, highlighting the obstacles that still lay ahead—namely data literacy and infrastructure—and how the philanthropic sector has an important role in shaping a fair digital ecosystem that is universally accessible and leaves no one behind.

It is only when a collaborative environment is created that it is possible to obtain that open sharing of information and Data. The sharing of knowledge, which, to an extent, is already an element of the international scientific community, unfortunately is not yet a standard element in this globalized world (for a question of technical, economic, political and legal obstacles).

It then becomes fundamental in this context to move from a siloed approach, to a cross sectorial culture of Data sharing, for which a possible solution lies in the creation of a sustainable model for private Data sharing, as Alberto Alemanno explains in his chapter, with the private sector in collaboration with philanthropic actors, using approaches such as Data stewardship as a means of responsibility toward the Data.

Through my role as Chair at the EFC in Brussels, I have had a great opportunity to introduce and stimulate this discussion, advocating it through dedicated, and increasingly appreciated, sessions during the annual EFC conferences where we were able to address the topic of Data at the service of philanthropy, and discuss the role of philanthropy in the creation of a Data Strategy, a Data market, sharing the experience of international experts in the field (*e.g. UNICEF, GovLab, Telefonica, Abeona Foundation, The Good Lobby, European Commission, GiveDirectly, The Data Guild, NESTA*).

This vision, reinforced since my arrival at Fondazione CRT, also reflects the longstanding investment of the Foundation in the field of Data science and complex systems through its support to ISI Foundation [8]: a forward-thinking action that was made in the conviction of the need to encourage the creation of a space in which the field of Data science and complex systems could grow. Over the course of this collaboration with Fondazione ISI, we have invested about 44 million euros through the Lagrange Project, providing over 800 young researchers with scholarships and research grants for projects on complex systems science and Data science, aiming to consolidate the links between academia, research, and the business world.

In recent years, this collaboration has been further strengthened around the topics of Data for good and impact. It is no surprise that the winner of the Lagrange Prize [9] in 2018 was César A. Hidalgo [10], for his work on understanding how teams, organizations, cities, and nations learn. A cross over between research areas that was also seen in the work of John Brownstein [11], the 2016 recipient of the Award, who was recognized for his groundbreaking studies on digital health and global infectious disease surveillance.

By virtue of our experience matured through collaboration with ISI Foundation, a launch pad for a higher-level ambition was created for Fondazione CRT. In synergy with ISI Foundation, this ambition takes the form of the European Data Centre for Impact and Social Good, the Centre, within the OGR [12] in Turin, a knowledge hub targeting the challenge of using Data Science and AI to better identify needs, guide interventions, and evaluate the impact of decisions and policies.

Whilst the role and function of the Centre is to facilitate problem scoping, research, advocacy, networking and training, its main areas of focus consist of the Science of Success for Philanthropy, including global issues such as the Future of Cities, Migration and Internal Displacement, Impact Investing and Impact finance.

The next objective we want to fulfill is to overcome the common "Data to insight" approach, and move towards "insight to action", permitting functional access to relevant Data, especially private big Data sources, through the use of Data Collaboratives—public–private partnerships for the exchange and analysis of Data, a topic that Stefaan Verhulst explores thoroughly in his chapter. A key aspect of success for the Centre, which has been active for about a year now, will be precisely in its capacity to strengthen our partnerships with philanthropic institutions, nonprofit organizations, as well as with research centers and private Data owners to experiment with Data practices and impact change at scale.

The aforementioned autonomy of philanthropic institutions also consents this sector to make long-term strategies and use patient capital. These are essential elements to create space for experimentation, for the development of new models, safe spaces and frameworks where collaborations can take shape and where Data sharing is made available by Data owners—who, in a way, can be seen as Data investors for social good—for the benefit of impact organizations and ultimately of society.

References

1. Fondazione CRT—Cassa di Risparmio di Torino is a private non-profit organisation founded in 1991, based in the north west of Italy. it is Italy's third-largest banking foundation by assets.

It has distributed €1.9 billion in resources to the community, enabling more than 40,000 initiatives.

2. European Foundation Centre is a non-profit international organization based in Brussels that unites 250 members of institutional philanthropy from over 30 countries.

3. Three examples include the August 2019 declaration of intent from the Business Roundtable—the association that brings together the "number one" of 200 major American companies—on corporate responsibility, with a commitment to create value for wider groups of beneficiaries than just shareholders; the "open letter" (January 2020) from BlackRock CEO Larry Fink "A Fundamental Reshaping of Finance" emphasizing the responsibility that companies have a responsibility towards communities, and their role in addressing environmental (e.g. climate change) and social issues; as well as the launch of Jeff Bezos' Earth Fund in February 2020 which foresees $10 billion to fight climate change.

4. C. Fiennes, We need a science of philanthropy, in *Nature* **546**, 7657 (2017), https://doi.org/10.1038/546187a

5. L. T. McGill, Number of Registered Public Benefit Foundations in Europe Exceeds 147,000 (Donors and Foundations Networks in Europe, 2016), https://dafne-online.eu/wp-content/uploads/2019/08/pbf-report-2016-9-30-16.pdf

6. Giving USA 2020: charitable giving showed solid growth, climbing to $449.6 billion in 2019.

7. S. Verhulst, The Value of data Collaborative for Good.

8. ISI Foundation is a 36-year-old non-profit research institute. Its research focuses on Complex Systems, Data Science, Artificial Intelligence and their applications to Public Health and Social Impact, with international visibility in data-intensive domains such as mobility and epidemic modelling.

9. The Lagrange Prize is an annual International award for scientific research in the field of complexity sciences, its applications and dissemination, created by Fondazione CRT with the scientific coordination of the ISI Foundation.

10. César A. Hidalgo leads the Collective Learning group at The MIT Media Lab and is an Associate Professor of Media Arts and Sciences at MIT.

11. John Brownstein is a Professor of Medicine at the Harvard Medical School and the Chief Innovation Office at Boston Children's Hospital.

12. OGR: ogrtorino.it is a former large industrial building (about 360 k ft2) in Turin recently reconverted into an innovative and experimental centre for contemporary culture, art, research and start-up in partnership with US and EU accelerators.

The Value of Data and Data Collaboratives for Good: A Roadmap for Philanthropies to Facilitate Systems Change Through Data

Stefaan G. Verhulst

Introduction

We live in challenging times. From climate change to economic inequality and forced migration, the difficulties confronting decision makers are unprecedented in their variety, and also in their complexity and urgency. Our standard toolkit for solving problems seems stale and ineffective while existing societal institutions are increasingly outdated and distrusted.

These are cross-sectoral challenges affecting government, corporations and civil society. Philanthropy is certainly not immune. Modern philanthropy faces a range of distinctively 21st century difficulties. Increasingly, it is clear, we need not only new solutions but also new methods for arriving at solutions [1]. There is a profound need to reimagine how philanthropic decisions are made and how services and programs are designed and delivered. In particular, much as organizations in other sectors have leveraged new technologies to enhance their role and achieve their missions better, so too do we need to consider how philanthropic organizations can be made more innovative, legitimate and effective in the digital era, and leverage all the 21st century tools to meet 21st century challenges. So far, philanthropies seem to have lagged behind or have not been sufficiently incentivized to change.

The choices confronting us are not political, not about selecting from among particular ideologies or regimes. We need to replenish and innovate our toolkit of solutions and approaches from across the ideological spectrum, and we have to be open to all possibilities. The goal is fundamentally to rethink the means by which

The author would like to thank Andrew Zahuranec and Michelle Winowatan for their research assistance and Claudia Juech and Rachel Rank for their input to earlier drafts.

S. G. Verhulst (✉)
Co-Founder and Chief of R&D of the GovLab, NYU Tandon School of Engineering, New York City, USA
e-mail: stefaan@thegovlab.org

philanthropy can better achieve its mission and, in the process, help solve complex public problems.

One thing is clear: the type of broad systemic change that is necessary will not be easy, nor will it be achieved overnight. Nonetheless, we believe that such change is possible, and, in this paper, we argue that the road to a 21st century philanthropy runs through data and increased use of data—what some have called a new "digital civil society" [2].

Data poses its own challenges and risks, of course; the datafication of our era is not an unqualified good. We discuss some of its risks in Sect. 1. In the remaining sections of the paper we hope to show how philanthropy can harness the best possibilities of data while minimizing its risks. In particular, in Sect. 4, we argue for greater use of data collaboratives: a new form of public-private partnership that has the potential to transform the relationship between the philanthropic and private sectors, and in the process lead to genuine positive social transformation. Section 5 presents some key challenges to achieving that transformation, and Sect. 6, the conclusion, summarizes our key recommendations; it offers a roadmap for the positive and collaborative use of data in the philanthropic project.

The Importance of Datafication

Before examining the role of data in the philanthropic sector, it is good to consider the role of data more generally. It has become a cliché (yet also a truism) to say that we live in a data age, marked by a surge in data collection and storage that has resulted from exponential increases in networked computing and the rise of mobile phones and other digital devices with embedded processors (the so-called "Internet of Things"). The term *datafication* has gained currency in recent years to describe this phenomenon. It encompasses more widely used concepts like "big data" and "open data," and is evidenced, among other ways, in the rapid quantification of politics, economics, culture, and virtually every other sphere of human existence.

Combined with advances in data and computer science, the potential impact of this phenomenon is dramatic and disruptive. Datafication is sometimes compared to the introduction of the printing press by Gutenberg, which transformed the public sphere and gave rise to capitalism and democracy [3]; or, alternatively, to the invention of the microscope by van Leeuwenhoek, which transformed discovery and the process of measurement, thus unleashing tremendous innovation [4].

Datafication offers tremendous opportunities, including for philanthropy, some of which we discuss below. But it is important to acknowledge at the outset that it has also created new challenges and poses new types of risks. The Cambridge Analytica scandal of 2018 spurred increased scrutiny of the business models employed by many large technology companies at the center of the datafication movement. Questions about personal privacy, disinformation, and information security abound amid the rise of so-called "surveillance capitalism," whereby companies leverage and monetize individuals' personal and behavioural data in exchange for free services [5]. Other emergent uses of data and algorithms—such as predictive policing and automated social service delivery systems—have been shown to exhibit biases and exacerbate forms of inequality.

These and other data challenges have contributed to thoughtful considerations of how to mitigate risks and bolster individuals' data rights [6]. Less helpfully, we are witnessing the continued rise of what we could call the "destruction" narrative. Data has become a toxic concept among many civil society organizations (including philanthropies) and other stakeholders [7]. As result, the burden of proof for those who want to show the positive impact of data has become substantially higher than for those who would warn of data's risks.

What's clearly required is not an outright rejection of data, but a new framework that puts ethics, the public interest and responsible data handling at the heart of its approach [8]. There is some urgency to developing and articulating this framework. We exist at a moment of great opportunity, but we also face a danger of backsliding, in which legitimate concerns over privacy and individual rights are illegitimately used to justify roll backs in advances that have been made in fields like open data and transparency over the years.

It is this productive middle ground (acknowledging the risks, but also embracing the opportunities) that this paper attempts to stake out. In what follows, I examine how societal institutions, in particular those from the philanthropic sector, could engage in a more sophisticated and more considered—and more *informed*—examination of the pros and cons of datafication. The overriding objective, the core mission, remains constant: to maximize the public good while limiting harms [9].

The Value of Data for Good

Evidence suggests the possibilities that data offers both to donors and recipients of philanthropic funds. As an article in the *Wall Street Journal* put it: "Just as the ability to access and analyze mountains of new information is transforming corporations, the era of big data has the potential to revolutionize the world of philanthropy—for both donors and charities" [10]. However, much of the existing evidence for this proposition is anecdotal. The truth is that leveraging data for good is still a new phenomenon, and we are only beginning to really understand how and under which conditions it works best.

Four Value Propositions

In an effort to systematize some of the anecdotal evidence, we recently analyzed a number of examples and case studies in which data was used within the philanthropic and nonprofit sector. We found four broad ways in which data can make a difference. We discuss these categories below, along with some key examples.

(1) **Situational Awareness, Nowcasting and Response**

Time and time again, we have seen that data can help us better understand demographic trends, public sentiment, and the geographic distribution of various phenomena. In doing so, data contributes to improved situational awareness and response, helping philanthropies better fulfill their missions. The following two examples are illustrative:

- The National Association of State Workforce Agencies (NASWA) used data from the LinkedIn Economic Graph—which uses data from 590 million members, 30 million companies, 20 million open jobs, and 84 thousand schools [11]—to better understand workforce trends, including in-demand jobs and talent availability in given regions. This data allowed NASWA to advance policies that match the supply of unemployed workers with demand [12].
- The National Eating Disorder Association and the National Suicide Prevention Lifeline partnered with Facebook to help them detect attempted suicides and more quickly alert first responders. In this partnership, first responders receive wellness checks flagged by Facebook's pattern recognition tools that automatically log and flag posts and videos suggesting suicidal ideation [13].

(2) **Knowledge Creation through Better Cause and Effect Analysis**

The second main way in which data can help philanthropies is by creating a better understanding of linkages between otherwise unassociated phenomenon. In this use case, datasets dispersed across sectors can be combined and analyzed to determine cause and effect. This has a number of benefits, not least of which is that it helps ensure that those responsible for solving problems have insight into the root causes of the problems they are trying to solve. Examples include:

- New Philanthropy Capital's Data Labs project allows charities to send data about their beneficiaries to government analysts to match them with similar individuals or groups. These matches of government administrative data and private social-sector data allow charities to understand the impact of their interventions on the populations they hope to help [14].
- Ajah's research tool Fundtracker uses data obtained from federal tax returns, government disclosures, and corporate social responsibility reports to provide detailed giving profiles of Canadian foundations and corporations [15]. This identification of who funds whom, how often, and when allows donors to understand how groups spend their money and what goals they achieve [16].
- The Center for Better Aging uses national statistics, interviews and workshops with the public, and recent scholarship to inform its reports on the challenges faced by the United Kingdom's aging population. One recent report, for example, drew on a partnership with the Greater Manchester Combined Authority to identify the causes and consequences of social, economic, and health inequalities in the Greater Manchester area [17].

(3) **Prediction and Forecasting**

The third way in which data can help philanthropy is by improving predictive and forecasting abilities. New predictive capabilities enabled by the analysis of previously inaccessible datasets can help institutions be more proactive in their responses to crises, and also to avert crises before they occur. We have seen numerous cases where data has been used for prediction and forecasting, including:

- In partnership with Chemonics, Arizona State University has developed a tool that helps identify and counter potential extremist sentiment in Libya. The tool, called LookingGlass, scrapes social media data to map networks of perception, influence, and belief online, thus harnessing the power of big data to counter violent extremism [18].
- The International Center for Tropical Agriculture (CIAT) partnered with the Colombian Ministry of Agriculture and Clima y Sector Agropecuario Colombiano (Colombian Climate and Agricultural Sector; CSAC) to provide meteorological data to farmers along with data on the economics and agronomy of rice cultivation. This information helped farmers anticipate and respond to extreme or harsh climatic events such as droughts or floods [19].
- NetHope, by acquiring data from the private, public and humanitarian organizations, mapped the trajectory of new Ebola outbreaks in West Africa, thus helping to identify where further outbreaks might occur and better prevent further spread of the virus [20].

(4) **Impact Assessment, Evaluation and Experimentation**

Finally, increased access to data can help with monitoring and evaluating the real-world impacts of policies and interventions. This helps design better, more responsive and evidence based policies or services, and enables a process of iteration and constant experimentation. Examples include:

- UN Women studied how big data sources—specifically data from Twitter, Facebook, and radio—could augment evaluations of gender equality and women's empowerment initiatives. The group conducted sentiment analysis, mapped engagement over time, and searched for statistical co-occurrences on two projects in Mexico and Pakistan to achieve this goal [21].
- UN Global Pulse joined with the multinational Spanish banking group BBVA to better understand disaster resilience. The partners used anonymized sale payments and ATM cash withdrawal data from over 100,000 of BBVA's clients in Mexico during 2014's Hurricane Odile. These datasets allowed the parties to measure economic activity before, during, and after the hurricane made landfall to understand economic recovery time [22].
- The New York-based nonprofit DoSomething.org used survey data from the organization Power to Decide to inform their "Pregnancy Text" campaign design, performance indicators, and hypothesis [23].
- In 2014, United Kingdom government body Sport England spearheaded the "This Girl Can" campaign to reduce the gender gap in sports participation. The campaign focused heavily on social media and analyzed over 10 million social media posts

about sports and exercise by women to understand the issues facing them. Their analysis revealed that young women were discouraged from exercise because of concerns about their appearance while mothers felt guilty about not spending time with their family. Still other women felt they were not good enough [24].

The Value of Data to Philanthropies

Philanthropies can subsequently leverage these three ways of how data generates value toward (Fig. 1):

- Improving their overall operations—combining digital transformation tools and data analytics to making them run more efficiently and effectively (by for instance streamlining grants management but also becoming more evidence based by extracting insights from the data within the grants management system).
- Transforming the way they are governed, making them more accountable and transparent—a topic of major importance in the current philanthropic landscape.

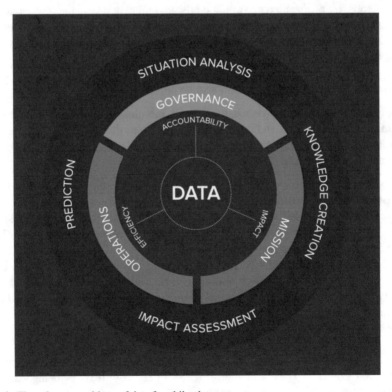

Fig. 1 The value propositions of data for philanthropy

- Increasing their impact—improving how they achieve their mission by making more evidence-based decisions and continuously adjusting their activities to take account of realities on the ground.

As such, when used responsibly and systematically, data can become a real differentiator in program effectiveness and impact for philanthropic organizations.

Three Different Data Sources

The above examples suggest that data has potential to transform how philanthropies go about achieving their mission. However, despite increasing recognition of data's potential and a growing effort to use it, our observation is that that potential remains often unmet because poor data literacy within the philanthropic world. There are cultural inhibitions toward becoming more evidence-based but also a lack of access to or appreciation of particular data sources.

In our work, we have identified three main data sources that could be relevant for philanthropies to consider (in addition to their own data assets) to further the use of data for good:

(1) **Open Government Data**

Over the past decade, a rapidly growing number of governments around the world have begun opening their data to the public, as a way to enhance transparency and accountability in governance. The US government's open data portal alone offers more than 120,000 publicly available data sets [25]. All this data offers tremendous possibilities for philanthropic (and other) organizations. Some ways in which they have utilized government data to advance their missions:

- JailVizNYC, an app created by Vera Institute, is a digital dashboard that provides data visualization using "Daily Inmates in Custody" data provided by the New York City government. This dashboard informs users of who is in jail, why they are there, and helps inform strategies to reduce the city's jail population [26].
- The Moldovan Ministry of Education has created the website AflaMD, which contains data on educational institutions in Moldova. The aim is to create transparency in Moldova's education budget by opening spending data for all schools in the country [27].
- Aclímate Colombia was created by actors involved in civil society, the private sector, and government to help independent farmers respond to the shifting weather patterns caused by climate change. The project uses open government datasets and other resources to "identify the most productive rice varieties and planting times for specific sites and seasonal forecasts" [28]. Researchers estimate the information could raise yield by up to 3 ha. [29].
- Tanzania's Education Open Data Dashboard and Shule are both online portals created to improve transparency of the country's education sector. Though the sophistication and detail provided varies, the platforms publish data on examination pass rates, pupil-teacher ratios, regional and district rankings, and other

statistics. Despite the low internet-penetration rate, the data can be used by civil society organizations to help families make better decisions about their child's education [30].

- In 2011, Slovakia passed legislation to require all documents related to public procurement be published online. The information from over 2 million contracts is used by the public, journalists, civil society groups, and international organizations to prevent corruption and make the government more transparent, responsive, and efficient [31].

(2) **Crowdsourced Data**

Technological innovation has dramatically changed the way people engage with each other and society in the digital world. In particular, social network platform and apps not only connect people on an individual level, but also facilitate new ways of sharing and collecting views, ideas and expertise or knowledge. So-called crowdsourced data is a powerful mechanism for philanthropies to harness the power of collective intelligence. It can help them in a variety of missions and tasks, most generally by tapping dispersed expertise to solve complex public problems. Some examples of how nonprofits are using crowdsourced data:

- Crowdsource Rescue connects affected people with civilian rescue volunteers in disaster areas through a crowdsourced map. Using their online platform, both individual rescuers and rescuees have the ability to tag their location on the map, making crisis response more rapid and efficient. The project started during Hurricane Harvey, in 2017, and has since been used for Hurricane Irma, Hurricane Nate, Mexico City Earthquakes, and Hurricane Maria [32].
- OurBrainBank is an app created by a nonprofit organization of the same name and used by Glioblastoma (a type of brain tumor) sufferers to log their symptoms. The aggregated information is analyzed by a team of medical researchers to deepen understanding of the disease, spot early signs of symptom change, help drive new trials, and develop new insights into the quality of life and experiences of people living with Glioblastoma [33].
- Through Google Waze's Connected Citizens program, The Society for the Protection of Nature in Israel leveraged Waze data to learn where they should focus their efforts on protecting wildlife [34]. The traffic data informs the organization of where roadkills usually happen, helping them become more efficient in patrolling and rescuing protected wildlife.

(3) **Privately Held Data (Notably Through Data Collaborative)**

A significant portion of data that we produce is held by private companies, such as mobile operators, social media platforms, and credit card companies. Historically, this data has been held in proprietary formats or silos, inaccessible to public use.

Yet as we have shown earlier in this piece, this situation is changing: A growing amount of private data is now being made available externally, notably to researchers and non-profit organizations, and this combination of private data and public access has tremendous potential for positive social transformation [35]. We call such combinations "data collaboratives;" we describe them further below, and we have written extensively about them elsewhere [36]. First, we consider some examples of how philanthropies are using private data to further their mission:

- The United Kingdom's Consumer Data Research Center collects various data, such as sales, consumer loyalty and rewards program, and product turnover, from retail and service businesses and uses this data for a broad range of study and economic policy to support innovation and improve governance and livelihood in general This information is available to researchers in criminology, health, transportation services, and other fields [37].
- The Global Fishing Watch is a project launched by the ocean conversation and advocacy group Oceana, the environmental watchdog Skytruth, and Google to help governments and the public monitor commercial fishing activity worldwide in almost real time. The system relies on Google's satellite data. It has since been adopted by the governments of Kiribati and Indonesia to enforce conservation policy and crack down on illegal fishing [38].
- Turk Telekom initiated a challenge, called Data for Refugees, to tackle migration issue in Turkey, particularly on how to improve the livelihood of Syrian refugees currently residing in Turkey. The company opened up access to three different datasets to selected projects. Among the data shared are location data and CDRs [39].
- 360Giving is a nonprofit and a registered charity in the United Kingdom that encourages grantmakers to publish information on "who, where and what they fund in an open standarised format" [40]. The charity wants to open up this information because it allows both funding organizations to make more informed decisions and fund seekers to draft better applications, boosting the impact of grant work [41].

The Importance of Data Collaboratives

As noted above, data collaboratives harbor particular potential for the use of *privately* held data in service of the *public* good—including corporate data but also data and insight held by the non-profit or academic sector. Indeed, if there is one lesson we have learned over the GovLab's six years of research, it is regarding the crucial importance of collaboration and partnerships in the data space. Over and over, in project after project, we have seen that greater collaboration between data users (demand) and data holders (supply), as well as among various sectors and domains of expertise, increases the real-world positive impact of data [42].

Collaboration also increases trust and ethics in the way data is handled and, importantly, the perceived legitimacy of such efforts [43]. In other words, data collaboration is important for philanthropic organizations looking to unlocking the potential of data and data science for good while limiting its risks and potential harms.

(1) What are Data Collaboratives?

The term "data collaboratives" refers to an emergent form of public-private part-nership in which actors from different sectors exchange and analyze data (and/or provide data science insights and expertise) to create new public value and generate new insights [44]. The potential and realized contributions of data collaboratives stem from the fact that the supply of and demand for data are generally widely dispersed—spread across government, the private sector, and civil society—and often poorly matched. Indeed, while most commentary on the data era's shortcomings focuses on the issues of data glut or misuse of data (e.g. through privacy violations) or, more recently, algorithmic bias, one of the biggest problems actually lies in a persistent failure to match supply and demand [45]. This failure results in tremendous ineffi-ciencies and lost potential. Much data that is released is never used; and much data that is actually needed is never made accessible to those who could productively put it to use.

Data collaboratives, when designed responsibly, are the key to addressing this shortcoming. They draw together otherwise siloed data and a dispersed range of expertise, helping match supply and demand, and ensuring that the correct institutions and individuals are using and analyzing data in ways that maximize the possibility of new, innovative social solutions [46].

(2) Examples of Data Collaboratives

At the GovLab, we have created a "Data Collaboratives Explorer" that contains more than 150 examples of collaboratives in geographical settings as varied as Nigeria, the United States, Bangladesh, and Denmark [47]. These examples show how private data has been used to solve public problems in sectors ranging from agriculture to public health to migration. A few examples:

- Esoko is a for-profit company that collects and pools data from governments and private sources for smallholder farmers across Africa. The data helps these farmers navigate complex global supply chains and improve their bargaining power against large, global actors [48].
- The Orange Data for Development Telecom Challenge was a competition hosted by the French multinational telecommunications company Orange to see how corporate data could be used to solve various development problems in the Ivory Coast and Senegal. In consultation with government ministries and inter-national experts, Orange provided anonymized, aggregated call detail record data to researchers. The winning research team used the mobile phone data for electrification planning [49].
- OpenTraffic, a nonprofit sponsored by the World Bank and the companies Conveyal and Mapzen, collects and shares accurate, real-time traffic data with individuals and organizations involved with transportation across the world. Its data sharing helps communities monitor roadway conditions and improve their transportation systems and make better planning decisions [50].
- In Chicago, a data collaborative helped public agencies, newsrooms, academics, and researchers better understand the local criminal justice system through

the sharing of arrest and investigatory stop data, snapshots of the county jail population, and information on the State's Attorney's cases [51].

(3) **Taxonomy of Data Collaboratives**

Out of the hundreds of examples we studied, we were able to identify six broad types of data collaboratives [52]. These are worth briefly considering because they provide templates for philanthropies (and other organizations) considering the use of private data for public good.

- **Research partnerships**, in which corporations share data with universities and other research organizations. Through partnerships with corporate data providers, several researchers organizations are conducting experiments using anonymized and aggregated samples of consumer datasets and other sources of data to analyze social trends. For instance: Yelp shares its data on neighborhood businesses with 30 universities for researchers to build tools and discover meaningful value in the data. Using shared data on Yelp businesses in the San Francisco Bay Area, an academic research team from U.C. Berkeley used a probabilistic model for natural language processing to detect subtopics across a dataset of over 200,000 Yelp business reviews. Their research uncovered correlations between positive ratings and service quality, giving business owners evidence for improving their services [53].
- **Prizes and challenges**, in which companies make data available to qualified applicants who compete to develop new apps or discover innovative uses for the data. Companies typically host these contests in an effort to incentivize a wide range of civic hackers, pro-bono data scientists and other expert users to find innovative solutions with the available data. For instance: Spain's regional bank BBVA hosted a contest inviting developers to create applications, services and content based on anonymous card transaction data. The first prize went, for instance, an application called Qkly, which helps users plan their time by estimating what time of day a given place will be most overcrowded so as to avoid lines [54].
- **Trusted intermediaries**, where companies share data with a limited number of known partners. Companies generally share data with these entities for data analysis and modeling, as well as other value chain activities. For instance: South Africa-based telecom MTN made anonymized call records available to researchers through a trusted intermediary, Real Impacts Analytics (Now Dahlberg Data)— data analytics firm that provides guided and predictive analytics solutions [55].
- **Application programming interfaces (APIs)**, which allow developers and others to access data for testing, product development, and data analytics. By signing a terms of service agreement, companies give access to streams of its data in order to build applications. For instance: Through its metadata and click tracking functionality, Bitly estimated social trends and allows users to build tools from real-time data [56].

- **Intelligence products**, where companies share (often aggregated) data that provides general insight into market conditions, customer demographic information, or other broad trends. For instance: Google shared search query-based data in conjunction with data from the US Centers for Disease Control in order to estimate levels of influenza activity over time [57].
- **Data cooperatives or pooling**, in which corporations —and other important data-holders such as government agencies— group together to create "collaborative databases" with shared data resources. These collaborations typically require an organizing partner as well as technical and legal frameworks surrounding the use and distribution of the data. For instance: Through its Accelerating Medicines Partnership, the US National Institutes of Health (NIH) is helping organize data pooling among the world's largest biopharmaceutical companies in order to identify promising drug and diagnostic targets for Alzheimer's disease [58].

Challenges associated with Data Collaboratives

The preceding discussion shows the tremendous potential offered by data for good through data collaboratives. That potential is real, demonstrated by a number of examples and, increasingly, recognized by philanthropies themselves. Yet, as we also noted at the outset of this paper, there are some equally real challenges associated with data collaborations and these need to be acknowledged so that they can be addressed. The ultimate goal is to create an ethical framework for using data for good that is effective, scalable, sustainable and responsible.

In this section, we focus on **seven** challenges:

- **Lack of Awareness and Data Literacy**: Both among those supplying data and those who might use it within philanthropies, there is a general lack of awareness and appreciation regarding the potential and value of how data can be used for the public good. This lack of awareness and data literacy limits interest in, and usage of, data.
- **Absence of Trust**: The field of data is characterized by a pervasive absence of trust. This is true between potential partners (e.g. the private sector and philanthropic data users) and also among the public at large, which remains ambivalent and skeptical about how its data is being used (and misused). While such concerns are understandable and often valid, the absence of trust acts as a barrier to the potential of data—and, importantly, suggests the need for a framework that would include guidelines and rules about how data is used by the philanthropic sector. Such a framework could help build trust, especially if it is made publicly available and is accompanied by robust steps for monitoring and to ensure accountability. This framework would necessarily need to start with responsibly managing the data assets philanthropies already possess, collected directly or by their grantees, before considering external data sources.

- **Private Sector Uncertainties**: The previous section has outlined the tremendous potential of private sector data. Yet companies often have concerns and reservations about the reuse of their data. While many of these concerns are legitimate, they too act as barriers on the philanthropic use of data. An (incomplete) list of concerns that we have encountered in our research includes the following:

  - Data leaks and competitors gaining business intelligence about markets and operations;
  - Penalties and fines from regulators or other lawmakers imposed due to the interpretation of legislation and processes;
  - Reputational loss if customers grow suspicious of how their data is being used and recycled.
  - In our experience, while these reservations often go unstated, they are the implicit reasons why companies do not allow fuller access to their data or participate more actively in data collaboratives.

- **Limited Capacity**: The ability to process, analyze and use big data varies widely by organization, another factor which limits its potential positive public impact. This lack of capacity is true across sectors, but it is perhaps particularly pronounced among nonprofit organizations, many of which are have failed to invest in staff with data skills or only now adapting to the new digital era in which they operate. Furthermore, as noted earlier, capacity limitations are a particular problem for smaller philanthropies, which are often those operating on the frontlines and with most direct exposure to the problems that need solving.

- **Transaction Costs**: While private-sector data is often (though not always) made available without charge, it would be incorrect to assert that opening up and using data is always free. Transaction costs are incurred at several points in the data life-cycle—while preparing data; de-risking data (e.g. through anonymization); and in coordinating with partners, including through the preparation of legal agreements or other structures, mechanism or institutions to permit data sharing and reuse. Our research indicates that such costs are especially burdensome, once again, for smaller philanthropies, and in the context of the developing world.

- **Scaling Challenges**: In our experience, most philanthropic projects that use data are relatively small and limited in their impact because of the scale of implementation. Limiting factors to scale up pilots include lack of mechanisms as platforms to help identify partners or opportunities to collaborate and lack of industry-specific guidelines and protocols to guide collaborations. In developing countries, limited funding also hinders the scaling up successful test cases.

- **Limited Community of Practice and Knowledge Base**: Finally, the nascent nature of the field poses an additional barrier. Successful data projects require a community of practice and build upon a knowledge base (including, for example, case studies and lessons learned). These conditions are virtually absent in the philanthropic sector. However, as data initiatives continue to multiply we would expect to see the emergence of new bodies and institutions that could offer the foundations of such a community of practice and learning.

A Roadmap for Philanthropies to Facilitate Systems Change Through Data and Data Collaboratives

The value of data for good is starting to become apparent in diverse and ongoing contexts, but much remains to be done. The evidence base is thin, and there is an absence of a systemic, structured approach that can be replicated across projects and geographies. This absence is particularly true when it comes to demonstrating and understanding the benefits of data collaboration.

One of the aims of the above discussion has been to take first steps toward a more targeted and systemic approach. In conclusion, we recommend that philanthropies focus on the following six steps to maximize the potential of data in their work, all while increasing and enhancing the possibilities of data collaboration. Taken together, these steps can be considered a "Roadmap for Philanthropies to Facilitate Systems Change through Data and Data Collaboratives":

(1) **Increase Evidence and Awareness**

The fledgling and ill-defined nature of the field poses challenges to greater adoption of data in philanthropy, and in particular to the potential of data collaboratives. Simply put, if organizations and individuals don't know about the potential of data and collaboration, then their actual use and impact will be limited. In order to spur greater adoption of data and the data collaboration model, philanthropy needs to create and better document evidence of the potential value of collaboration and raise awareness among key target communities.

(2) **Increase Readiness and Capacity**

For all the evident benefits of data, many organizations continue to display a certain reluctance or reticence to share it. This reluctance is true when it comes to using data in general and especially true when it comes to data collaboration, where mistrust, misaligned incentives or priorities and other obstacles continue to impede progress. A lack of technical capacity is also a major obstacle to greater uptake of data, especially for smaller organizations that lack the type of specialized knowledge often necessary to store and analyze data.

For all these reasons, there is an urgent need to develop new capacities and a new readiness within philanthropic organizations. From top to bottom, organizations need to incorporate data into their daily operations, fostering data literate staff and board members and drafting data strategies. This development can be done with more emphasis on training and skill-building, as well as with greater cross-sectoral collaboration so that data specialists in the private sector, for instance, may contribute some of their skills to philanthropy. Some of these goals (though by no means all) may be aided by the awareness building mentioned above.

(3) **Address Data Supply and Demand Inefficiencies and Uncertainties**

There are two sides to the data collaboration equation: supply and demand. To amplify the benefits of collaboration we need to address inefficiencies and "market failures" on both sides. This involves mapping and auditing the data they already have in-house as part of their planning, due diligence and grantee reporting. It also includes proactively reaching out to potential supply-side organizations, especially in the private sector, and working with them to minimize concerns over competitiveness or reputation, and to define data responsibility approaches taking into account both the value proposition and potential risks of collaboration.

Equally, understanding the demand for data is a vital part of establishing a responsible and impactful data collaboration process. Yet here too, as on the supply side, there are a number of inefficiencies limiting impact. Many of these are related to readiness, capacity and awareness—issues which we have addressed above.

(4) **Establish a New "Data Stewards" Function**

We also believe that certain institutional changes are required to maximize the potential of data collaboration in philanthropy. In particular, we propose the establishment of a new role that would be embedded in any organization dealing or considering dealing with data: data stewards. Data stewards would be the individual (or individuals) within an organization responsible for setting data policy and for steering and encouraging collaborative approaches. Data stewards also play a central role in ensuring that data is shared and handled responsibly, to address some of the inherent risks of open data and data collaboration. We have written extensively about data stewards elsewhere [59]. Essentially, they would function as the linchpin of a new, more systemic framework for responsible data collaboration in philanthropy.

(5) **Develop and Strengthen Policies and Governance Practices for Data Collaboration**

Establishing the new role of data stewards is but one aspect of a broader task: establishing a clearer and better defined governance framework for data collaboration in the philanthropic sector. There is clear evidence across sectors that clearly articulated governance mechanisms, along with auditing and accountability mechanisms, enhance trust and the usage of data [60]. While some policies and mechanisms are universal and can be transposed to philanthropy, others may require sectoral adaptation and changes. The overall goal is to develop a well-defined set of guidelines and principles that covers the entire data lifecycle within the philanthropic sector.

(6) **Strengthen the Ecosystem**

Few successful initiatives (technical or otherwise) emerge in isolation. They are always supported, incubated and nurtured by a well-developed ecology that includes thought-leaders, funding, institutional and regulatory support, and a variety of other components. One of the challenges confronting data in philanthropy is that the current ecosystem is weak. There is a need to build capacities and networks of association where knowledge can be shared and funding sources identified. Likewise, there is a need for a robust body of evidence (case studies, examples, etc) that can provide lessons and best practices, and that can serve as the foundation of new projects and initiatives.

I am under no illusion that the six steps outlined above will be easy to implement, or that change will occur rapidly. What's needed, as we suggested at the beginning of this paper, is wholescale, systemic transformation, a transformation that will embed data and data collaboration into everyday philanthropic practice. Such a transformation (what we might call the datafication of philanthropy) is likely to come about slowly, and in an incremental fashion. Yet despite its difficulties, it remains central to the practice of philanthropy in the twenty-first century—and, because philanthropy itself is so important, its datafication is crucial to mitigating many of the most vexing problems we face as a society, polity and economy.

References

1. S. Verhulst, From #Resistance to #Reimagining Governance: 6 Shifts That Can Improve the Way We Solve Public Problems (OpenDemocracy, 2017), https://www.opendemocracy.net/ste faan-g-verhulst/from-resistance-to-reimagining-governance-6-shifts-that-can-improve-way-we-solve.
2. Stanford PACS, Digital Civil Society Lab, Center on Philantropy and Civil Society, https://pac scenter.stanford.edu/research/digital-civil-society-lab/.
3. B. Woodley, The Impact of Transformative Technologies on Governance: Some Lessons from History, at Institute on Governance (Ottawa, Canada, October 2001), https://www.files.ethz. ch/isn/121785/transformative_tech.pdf.
4. A. McAfee, E. Brynjolfsson, The Big Data Boom Is the Innovation Story of Our Time (The Atlantic, 2011). https://www.theatlantic.com/business/archive/2011/11/the-big-data-boom-is-the-innovation-story-of-our-time/248215/.
5. S. Zuboff, *The Age of Surveillance Capitalism: The Fight for a Human Future at the New Frontier of Power* (PublicAffairs, New York, 2019).
6. M. Tisne, It's Time for a Bill of Data Rights (MIT Technology Review, 2018). https://www.tec hnologyreview.com/s/612588/its-time-for-a-bill-of-data-rights/.
7. The toxic reputation of data among civil society organizations is the consequence of well-publicized failures by technology advocates to consider the public's low tolerance for risk and uncertainty or adequately protect beneficiaries from harm. For one notable example, see: M. Bulger, P. McCormick, M. Pitcan, The Legacy of InBloom (Data & Society, 2017), https://dat asociety.net/pubs/ecl/InBloom_feb_2017.pdf.
8. S. Verhulst, Data Responsibility: A New Social Good for the Information Age (The Conversation, 2017), http://theconversation.com/data-responsibility-a-new-social-good-for-the-inform ation-age-67417.
9. S. Verhulst, Corporate Social Responsibility for a Data Age (Stanford Social Innovation Review, 2017), https://ssir.org/articles/entry/corporate_social_responsibility_for_a_data_age.
10. L. Bernholz, How Big Data Will Change the Face of Philanthropy (The Wall Street Journal, 2013), https://www.wsj.com/articles/how-big-data-will-change-the-face-of-philanthropy-138 6882561.
11. LinkedIn Economic Graph (LinkedIn, 2019), https://economicgraph.linkedin.com/.

12. National Association of State Workforce Agencies, NASWA/LinkedIn Partnership (2018), https://www.naswa.org/partnerships/linkedin.
13. V. Callison-Burch, J. Guadagno, A. Davis, Building a Safer Community With New Suicide Prevention Tools (Facebook Newsroom, 2017), https://newsroom.fb.com/news/2017/03/bui lding-a-safer-community-with-new-suicide-prevention-tools/.
14. NPC, Data Labs, https://www.thinknpc.org/examples-of-our-work/initiatives-weve-worked-on/data-labs/.
15. Ajah Fundtracker Pro, http://www.fundtracker.ca/.
16. J. O'Kane, Using Public Data, Entrepreneurs Wring Cash from the World of Charitable Giving (The Globe and Mail, 2018), https://www.theglobeandmail.com/report-on-business/small-bus iness/sb-growth/the-challenge/using-public-data-entrepreneurs-wring-cash-from-the-world-of-charitable-giving/article15137868/.
17. Center for Ageing Better, Addressing Worklessness and Job Insecurity amongst People Aged 50 and over in Greater Manchester (Centre for Ageing Better, 2017). https://www.ageing-better.org.uk/publications/addressing-worklessness-and-job-insecurity-amongst-people-aged-50-and-over-greater.
18. M. Piercey, C. Forbes, H. Davulcu, Through the Looking Glass: Harnessing Big Data to Respond to Violent Extremism (Devex, 2016), https://www.devex.com/news/sponsored/thr ough-the-looking-glass-harnessing-big-data-to-respond-to-violent-extremism-88503.
19. S. Verhulst, A. Young, Climate Modeling in Colombia (Data Collaboratives, 2018), http://dat acollaboratives.org/cases/climate-modeling-in-colombia.html.
20. G. Olaffson, Coordinating Data Share For Ebola Outbreak In West Africa (NetHope Blog, 2014), https://nethope.org/2014/08/25/coordinating-data-share-for-ebola-outbreak-in-west-africa/.
21. C. A. Lopes, S. Bailur, G. Barton-Owen, Can Big Data Be Used for Evaluation? A UN Women Feasibility Study (UN Women Headquarters, 2018), http://www.unwomen.org/-/media/headquarters/attachments/sections/library/publications/2018/can-big-data-be-used-for-evaluation-en.pdf?la=en&vs=3013.
22. E. Martínez, M. Alfaro, R. Hernández, R. Maestre Martinez, J. Murillo Arias, D. Patane, A. Zerbe, R. Kirkpatrick, M. Luengo-Oroz, Measuring People's Economic Resilience to Natural Disasters (BBVA AI Factory), https://www.bbvadata.com/odile/.
23. B. Kanther, How DoSomething Uses Data to Change the World (Socialbrite, 2012), http://www.socialbrite.org/2012/10/15/how-dosomething-uses-data-to-change-the-world/.
24. S. Verhulst, A. Young, The Potential of Social Media—Intelligence to Improve People's Lives: Social Media Data for Good (The GovLab Report, 2017) https://papers.ssrn.com/abstract=314 1457.
25. Data.Gov, https://www.data.gov/.
26. NYC Open Data, Open Data. Connecting New Yorkers, https://opendata.cityofnewyork.us/.
27. Moldovan Ministry of Education—Schools Portal (Afla.MD, accessed April 2021), https://www.programmableweb.com/api/aflamd.
28. CGIAR and CCAFS, Big Data for Climate-Smart Agriculture (Research Program on Climate Change, Agriculture and Food Security, 2015), https://ccafs.cgiar.org/research/projects/big-data-climate-smart-agriculture.
29. CGIAR and CCAFS, Big Data for Climate-Smart Agriculture (Research Program on Climate Change, Agriculture and Food Security, 2015), https://ccafs.cgiar.org/research/projects/big-data-climate-smart-agriculture.
30. S. Verhulst, A. Young, J. McMurren, D. Sangokoya, Open Education Information In Tanzania (Open Data's Impact, 2016), http://odimpact.org/case-open-education-information-in-tanzania.html.
31. A. Clare, D. Sangokoya, S. Verhulst, A. Young, Open Contracting And Procurement In Slovakia (Open Data's Impact, 2016), http://odimpact.org/case-open-contracting-and-procurement-in-slovakia.html.
32. Crowdsource Rescue, https://crowdsourcerescue.com/.
33. Our Brain Bank, http://ourbrainbank.com/.

34. B. Thompson, Popular road kill reporting app saves wild lives (Green Prophet, 2017), https://greenprophet.com/2017/08/this-popular-app-helps-you-report-roadkill-to-save-the-wild/..

35. See also GovLab's listing of data collaboratives: Data Collaboratives Explorer (Data Collaboratives, 2018), http://datacollaboratives.org/explorer.html.

36. S. Verhulst, A. Young, Introduction (Data Collaboratives, 2018), http://datacollaboratives.org/introduction.html.

37. S. Verhulst, A. Young, Consumer Data Research Centre (Data Collaboratives, 2018), http://datacollaboratives.org/cases/consumer-data-research-centre.html.

38. B. Dennis, How Google Is Helping to Crack down on Illegal Fishing—from Space (Washington Post, 2016), https://www.washingtonpost.com/news/energy-environment/wp/2016/09/15/from-space-a-new-effort-to-crack-down-on-illegal-fishing-across-the-globe/.

39. Ali Salah et al., *Guide to Mobile Data Analytics in Refugee Scenarios: The 'Data for Refugees Challenge' Study* (Springer International Publishing, 2019)

40. 360Giving (2019), https://www.threesixtygiving.org/about/.

41. 360Giving (2019), https://www.threesixtygiving.org/about/.

42. M. Susha, S. Janssen, Verhulst, Data collaboratives as "bazaars"? A review of coordination problems and mechanisms to match demand for data with supply, in Transforming Government: People, Process and Policy **11**(1), 157–172 (2017). https://doi.org/10.1108/TG-01-2017-0007.

43. C. Ansell, A. Gash, Collaborative Governance in Theory and Practice in Journal of Public Administration Research and Theory **18**(4), 543–71 (2008). https://doi.org/10.1093/jopart/mum032.

44. S. Verhulst, A. Young, Introduction (Data Collaboratives, 2018), http://datacollaboratives.org/introduction.html.

45. S. Verhulst, I. Susha, A. Kostura, Data Collaboratives: Matching Demand with Supply of (Corporate) Data to Solve Public Problems (Medium, 2016), https://medium.com/@sverhulst/data-collaboratives-matching-demand-with-supply-of-corporate-data-to-solve-public-problems-dc75b4d683e1#.wq7bnrtue.

46. S. Verhulst, A. Young, Introduction (Data Collaboratives, 2018), http://datacollaboratives.org/introduction.html.

47. See also GovLab's listing of data collaboratives: Data Collaboratives Explorer (Data Collaboratives, 2018), http://datacollaboratives.org/explorer.html.

48. S. Verhulst, A. Young, Esoko (Data Collaboratives, 2018), http://datacollaboratives.org/cases/esoko.html.

49. S. Verhulst, A. Young, Orange Telecom Data for Development Challenge (Data Collaboratives, 2018), http://datacollaboratives.org/cases/orange-telecom-data-for-development-challenge-d4d.html.

50. Open Traffic, http://opentraffic.io.

51. Chicago Data Collaborative, https://chicagodatacollaborative.org/

52. S. Verhulst, A. Young, Introduction (Data Collaboratives, 2018), http://datacollaboratives.org/introduction.html.

53. UC Berkeley School of Information, Students' Data Analysis Uncovers Hidden Trends in Yelp Reviews (2013), https://www.ischool.berkeley.edu/news/2013/students-data-analysis-uncovers-hidden-trends-yelp-reviews.

54. M. A. Iñesta, BBVA on the Trail of Its Own Applications Ecosystem (BBVA Innovation Center, 2013), https://web.archive.org/web/20170713135408/http://www.centrodeinnovacionbbva.com/en/blogs/entrepreneurs/post/bbva-trail-its-own-applications-ecosystem.

55. S. Verhulst, Mapping the Next Frontier of Open Data: Corporate Data Sharing (Medium, 2015), https://medium.com/internet-monitor-2014-data-and-privacy/mapping-the-next-frontier-of-open-data-corporate-data-sharing-73b2143878d2.

56. S. Verhulst, Mapping the Next Frontier of Open Data: Corporate Data Sharing (Medium, 2015), https://medium.com/internet-monitor-2014-data-and-privacy/mapping-the-next-frontier-of-open-data-corporate-data-sharing-73b2143878d2.

57. JP Morgan Institute, Home page (accessed April 2021), https://www.jpmorganchase.com/institute.

58. S. Reardon, Pharma Firms Join NIH on Drug Development (Nature News, 2014), https://doi.org/10.1038/nature.2014.14672.
59. A. Young, About the Data Stewards Network (Data Stewards Network, 2018), https://medium.com/data-stewards-network/about-the-data-stewards-network-1cb9db0c0792.
60. C. Ansell, A. Gash, Collaborative Governance in Theory and Practice in Journal of Public Administration Research and Theory **18**(4), 543–71 (2008). https://doi.org/10.1093/jopart/mum032.

UN Global Pulse: A UN Innovation Initiative with a Multiplier Effect

Paula Hidalgo-Sanchis

Introduction

UN Global Pulse is recognized today worldwide as centre of excellence on the use of big data and artificial intelligence for the public good. It is a flagship innovation initiative on big data under the Executive Office of the United Nations (UN) Secretary-General that was launched with the mission to accelerate the discovery, development and scaled adoption of big data innovation for sustainable development practice and humanitarian action. In 10 years, UN Global Pulse has been an innovation catalyzer in the UN system, it ignited innovation not only in the UN system but also in government counterparts and partners across the innovation ecosystem.

The initiative is a great example on how a strategic venture can inspire innovation inside an organization and have a multiplier effect. As a stone thrown in the water, the initiate had a ripple effect inspiring new ways of working, producing ground breaking hi-tech tools, developing new methods, consolidating new partnerships and contributing to new ways of approaching development practice and humanitarian assistance. A relatively small but enthusiastic team of around 50 people spread across the globe working as a network on a research agenda based on experimentation yielded results and had impact beyond expectations. In this chapter, we unfold UN Global Pulse's strategy relating milestones achieved and describing the ripple effect of these milestones.

Dr. P. Hidalgo-Sanchis (✉)
Senior Programme Manager, United Nations Global Pulse, New York City, USA
e-mail: hidalgo-sanchis@unglobalpulse.org

© The Author(s), under exclusive license to Springer Nature Switzerland AG 2021
M. Lapucci and C. Cattuto (eds.), *Data Science for Social Good*,
SpringerBriefs in Complexity, https://doi.org/10.1007/978-3-030-78985-5_3

UN Global Pulse's Implementation Strategy

To achieve its mission, UN Global Pulse organized its work under a two-pillar implementation strategy (Fig. 1): Track 1 (Innovation Driver)—the implementation of projects to incorporate digital and big data to improve data-driven decisions; and Track 2 (Ecosystem Catalyst)—the implementation of solutions to strengthen the environment for the adoption of big data innovations and enabling frameworks.

With three innovation Labs in New York, Jakarta and Kampala, UN Global Pulse achieved progressively important of milestones that contributed to the recognition of the initiative worldwide. Here are some highlights on these milestones:

- Extensive research and inspiring projects in countries around the world conducted in collaboration with hundreds of stakeholders from a variety of sectors, including research groups, private sector companies, UN agencies, donor countries of international assistance, Governments, Foundations, International Organizations and civil society organizations proved that big data sources are available, reliable and accessible to contribute to sustainable development and humanitarian action.
- New partnerships with private sector companies for the reuse of big data for the public good were established for the first time by the UN system. For example, UN Global Pulse agreements with international companies as Tweeter and Crismon Hexagon.
- The implementation of projects lead to the in-house development of sophisticated hi-tech tools praised globally that gave UN Global Pulse credibility in the tech space [1]. In some cases, these tools were ground breaking worldwide innovations as in the case of the radio content analysis tool [2] that allows gaining insights from public radio discussions in real time.
- UN Global Pulse has worked extensively in Data Privacy and Digital Ethics, leading the privacy policy work in the UN system. The initiative mobilized partners for the development of policy frameworks for the adoption of big data and artificial

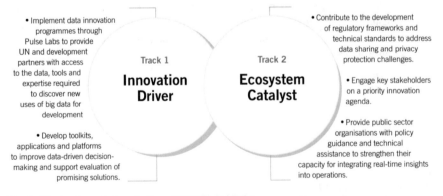

Fig. 1 The implementation strategy of UN Global Pulse

intelligence for the public good what lead to the first set of Data Privacy and Protection Principles adopted by all UN Agencies.

A visionary leadership, a global team of highly qualified and motived staff working as a network from 3 continents embracing a culture of collaboration are the ingredients of the recipe of the success of UN Global Pulse. To create the team, new skillsets were brought for the first time into the UN system in the hands of data engineers, data scientists, artificial intelligence advisors, data visualizations specialists, user experience designers, software developers, language analysts or transcription specialists. Communication and understanding between those professionals and more traditional UN staff as programme analysts, policy experts and operations mangers required building up bridges and maintaining an ongoing dialogue.

The research agenda based on experimentation had operational needs similar to those of startup companies and different to traditional UN projects. The UN system lacked the flexibility required to accommodate daily operations ruled by evolving and unplanned implementation needs. These are the three main challenges that the initiative encountered for implementation:

- Lengthy and rigid procurement processes that did not respond to the needs to develop technology products and affected directedly the team's abilities and timeframes for implementation.
- Recruitment of personnel was very complicated as human resources parameters were not adapted to technology environments. For example, the UN system required a minimum number of years of working experience for a certain contract grade, while the best candidate identified for a job held a Ph.D. in a very specialized area of work but had no working experience.
- New types of legal agreements were needed to access and analyse big data processed by private sector companies, to write these, the dialogue between legal units of telecoms and UN system took years.

The team at UN Global Pulse developed innovative new ways of working to overcome implementation challenges. To achieve implementation milestones, the management team at UN Global Pulse, formed by the Director, Deputy-Director, the Chief Data Scientist and the Managers of Pulse Lab Kampala and Pulse Lab Jakarta, navigated operational challenges with creativity and in many cases taking risks.

The success of UN Global Pulse is being measured based on demand to provide services, support to implement projects and produce tools. At the time of writing this chapter, the demand is so high that the small team cannot respond to it. On top of this demand from UN Agencies and partners, the initiative was asked to provide technical support the UN Secretary General Emerging Technology Lab and to support the UN Reform expanding it's network of Labs.

Pioneering New Areas of Work

Work conducted over 10 years with 17 UN agencies in partnership with private, public sector and academia showed that a variety of big data sources and artificial intelligence tools can support empowering vulnerable populations, protecting the environment or preventing violent extremism. The types of data, analysis and uses varied considerably. Records of purchases for example give new light related to consumption patterns that inform policies on non-communicable diseases. Mobile phone data in the aftermath of a natural disaster supports emergency response to communities. Photos of haze taken by citizens enhances monitoring of unfolding shocks to population groups. Work in 4 continents probed that predictive analytics support risk management, accountability and enhances response to vulnerable populations and the environment.

Exploring the effects of extremist violence on online hate speech

---

The increased prevalence of hate speech online can reinforce the subordination of targeted minorities. UN Global Pulse explored the potential causal relationships between violence committed offline—either in the name of Islam or intentionally directed against Muslim communities—and the fluctuation of derogatory, offensive, or even inciting comments online. Partners: IBM Science for Social Good, University of Pompeu Fabra (Spain).

The team at Global Pule implemented a large number of projects with many types of big data including anonymized mobile phone data. This data is processed by telecom companies and it is considered the richest source of big data because of the insights that it provides about societal behavior. UN Global Pulse conducted extensive work exploring the uses of mobile phone data for humanitarian assistance and sustainable development, some highlights of this pioneering work are presented in this section.

Using Call Detail Records (CDR) to Understand Refugee Integration in Turkey [3]

Integration is a complex and gradual legal, economic, social, and cultural process that burdens both the settling population and the receiving society. This research aimed to contribute to a new methodological framework to measure and evaluate integration through the lens of spatial and social segregation using anonymized CDR data. Partners: UNHCR and Turk Telecom.

Using Mobile Network Data to Inform Disaster Response in Asia

Covering the Highlands earthquake in Papua New Guinea and the Ambae Volcano in Vanuatu, Pulse Lab Jakarta developed insights on internal displacement to inform the targeting of humanitarian assistance. The analysis indicated that mobile network data be used to design evacuations routes in near real-time and to build predictive models for evacuee destinations. Partners: Digicel and Government of Indonesia.

Exploring the Potential of Mobile Money Transactions to Inform Policy

Mobile money is used in Uganda for domestic money transfers, to buy airtime and to pay bills, among others. The service brings banking facilities to communities traditionally unbanked. The analysis conducted showed correlations between subscriptions to mobile money services and patterns in social networks, which is useful to understand the profile of mobile money users and design programmes for its the expansion. Partners: Telecoms in Uganda.

Analysing seasonal mobility patterns using mobile phone data

Sudden changes in mobility patters can indicate shocks affecting vulnerable population groups as they might be a result of changes in livelihoods or coping strategies, or exposure to new. The research explored how monitoring such changes in real time could support early warning systems for informed decision-making and rapid response in Senegal. Partners: Orange, Politecnica University, UN World Food Programme.

Ripple Effect—Advancing Data Philanthropy and Collaboration with Private Sector

To conduct with anonymized CDRs, UN Global Pulse maintained a dialogue at global, regional and local level over the years with telecom companies and had to overcome many bottle necks. These bottle necks also apply to the work with other private sector companies. While some big data and digital data sources are publicly accessible on the open web, such as social media, valuable insights can be gleaned from data held by corporations ranging from telecommunications, to finance, food, pharmaceutical and transport. Although more of this data is being opened for research purposes, much remains to be done to create the conditions for sustainable access and use of these data for the SDGs. Some stakeholders around the world promote a *data philanthropy* approach, others argue that private sector has the right to sell data and others claim that data ultimately belong to the citizens that generate it, not to companies.

Along with the GovLab main challenges and barriers were mapped out as a starting point to define programmes to address them.

1. *Lack of Public Awareness*: There is a lack of awareness regarding the potential and value of privately processed data being deployed for the public good among those supplying data and those using it and among citizens.
2. *Absence of Trust*: The relationship between the private sector and governments or actors from civil society (including researchers) is often uneasy, and there is frequently a reluctance to collaborate and use the formers' assets. Citizens are also reluctant about private companies releasing their information even if it is anonymised.

3. *Private Sector Uncertainties*: An (incomplete) list of concerns encountered through our research include those concerning:

   - Data leaks and competitors gaining business intelligence;
   - Penalties and fines from regulators or other lawmakers imposed due to the interpretation of legislation and processes;
   - Reputation loss if customers grow suspicious of governments using their data for surveillance or other purposes;
   - Apprehensions over how public sector may use (or misuse) data on citizens.

4. *Limited Capacity*: The ability to process big data its limited among data suppliers, including lack of ability to anonymize data. On the demand side there is limited capacity to analyse large unstructured datasets.

5. *Transaction Costs*: The preparation of big data and establishing agreements for reuse imply costs.

6. *Scaling Challenges*: Limiting factors to scale up pilots with big data for the public good include lack of industry-specific guidelines and protocols to guide collaborations and limited funding.

7. *Limited Community of Practice and Expertise*: Given the nascent nature of the field, there is absence of a well-defined community of practice and expertise.

Building Evolving Technology Tools

With the implementation of projects to discover and tap into new data sets, UN Global Pulse developed in-house a wide range of technology toolkits or tools. The tools are based on different technologies, from data engineering to machine learning and artificial neural networks.

Tools built by Pulse Lab Jakarta [4]:

Haze Gazer
Haze Gazer enhances disaster management by providing real-time insights on the locations of fire and haze hotspots; the strength of haze in population centres; the locations of the most vulnerable cohorts of the population; and most importantly, the response strategies of affected populations. The tool uses advanced data analytics and data science to mine a variety data, from citizen's reports to satellite imagery.
DisasterMon
Integrated big data analytics and visualization tool to provide timely insights for natural disaster monitoring, emergency response and management of cyclones, earthquakes, hurricanes and floods in the Pacific region. Data feeds come from open data platforms, national statistics and social media.

(continued)

(continued)

VAMPIRE
The platform provides integrated map-based visualisations that show the extent of drought-affected areas, the impacts on markets, and the coping strategies and resilience of affected populations.

The technology tools evolved over the years with technical developments responding to different user needs and improvement of the digital applications. The team is now at the last stage of this organic process building global tools that can be customized to different users.

PulseSatellite

In Uganda, the type of roof of a dwelling, is used as a proxy-indicator for poverty by the Uganda Bureau of Statistics. Traditional thatched roofs harbor pests and disease and are high maintenance. And so as soon as someone can afford to, the roof will upgrade to a metal of tiled roof. At the same time, a wide variety of remote sensing image sources from highly light-sensitive satellites, high-resolution (< 1 m) image data, nano-satellites, balloon mapping, and low-cost unmanned aerial vehicles are available for measuring sustainable development. To support monitoring SDG1, Pulse Lab Kampala developed image processing software to count the roofs and identify the type of material they are constructed from.

Building on this work, UN Global Pulse started a collaboration with UNOSAT, a technology-intensive programme of the UN Institute for Training and Research (UNITAR). The team then developed PulseSatellite, a tool that allows automatizing the process of detecting structures from satellite imagery to support the production of maps and analytical reports that help humanitarian operations for migrant and refugee operations.

In dialogue with the Uganda Bureau of Statistics (UBOS), Pulse Lab Kampala is now customizing PulseSatellite to monitor in real time the extend and growth of slums in Kampala. This work is driven by the assumption that big data sources have the potential to complement traditional national statistics with real-time, targeted updates. In this context, the monitoring of the Global Goals presents an opportunity to explore the synergies between big data and traditional statistics and to incorporate new data sources and new methods by national statistical bureaus [5].

40% of Kampala's population lives in informal settlements predominately developed near wetlands throughout the city, without basic infrastructure such as water services, storm drainage, sewage treatment and solid waste collection. UBOS proposed to monitoring trends in the growth of the slums to support the provision of public services.

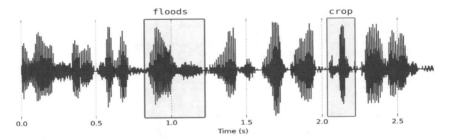

Fig. 2 Example of transcription from audio into text

Qatalog

Every day 7.5 million worlds are spoken in Uganda on radio shows. According to UNESCO, radio is the most reliable and affordable medium of accessing information and sharing in Africa. In Uganda, like many other countries in East Africa, community radio stations enable isolated communities to voice their concerns. Talk and phone-in shows are popular among all ages, on air, ordinary citizens discuss issues that are central to them [6]. Public radio discussions include first-hand reports of incidents that are not recorded elsewhere. The unconstrained nature of radio discussions, compared to other types of exchanges, means information collected includes perceptions that cannot be gathered from other sources.

Pulse Lab Kampala and partners build a deep learning system performing real-time data collection, speech recognition and content analysis of rural public talk radio programs in indigenous Ugandan languages (Luganda and Acholi). The system was later expanded to Somali language (Fig. 2).

To make the analysis accessible to the user, UN Global Pulse developed QataLog, a data mining tool that can extract, analyse and visualize data from different data sources. The first version of the tool allows extraction from social media and public radio. The tool uses a combination of optimized manual annotations techniques and automatic helpers that include translation, geo-location and text classification. The tool is being piloted with several UN teams with a very positive response and overwhelming demand.

Ripple Effect—Inspiring Global Platforms

Making accessible spatial data to inform biodiversity policy

Biodiversity is under threat globally, and policies are needed to preserve it. Because of the nature of biodiversity, planning exercises to identify, select and design areas to protect ecosystems and species, need to be based on spatial data. While a deluge of detailed spatial data (data geolocated in a geographic coordinate system) about the surface of Earth is available online and free of access from a variety of sources, this type of data is not always easily accessed and Geographical Information Systems (GIS) knowledge and tools are often required.

Pulse Lab Kampala developed the Nyanga Spatial tool [7] for the National Biodiversity Strategies and Action Plans (NBSAP) Forum. The geospatial tool, built on open source web technology, gave policy makers access to spatial data from a wide variety of sources allowing analysis and extraction of statistical summaries related for example to the extent of protected areas or changes in land use and climate. The tool design was inspired by others with similar functionalities as Global Forest Watch [8] and Protected Planet [9], both of which provide interactive online mapping and analytics for different types of environmental information. The tool was successfully tested to inform the National Biodiversity Plan II for Zimbabwe. Afterwards, UN Global Pulse guided the NBSAP to scale up the tool to become a global initiative. In 2018, the NBSAP launched the UN Biodiversity Lab [10] through a partnership between the UN Agencies (UNDP and UNEP) to provide countries with spatial data to make informed conservation decisions. By the time of writing this chapter (March 2019), 4 countries namely Ecuador, Haiti, Modova and South Africa had already used the new platform to inform biodiversity policy.

Policy Adoption: From Data Privacy and Protection to Ethics on AI

Big data and applied AI support evidence-based policies and programmes to advance SDG progress at an unprecedented scale. At the same time, the risks associated to the collection and processing of personal data at scale hinder harnessing its use for public benefit. Furthermore, the increasing implementation of AI creates an additional level of uncertainty in applying well settled and existent data protection norms as transparency, purpose specification, accountability or data minimization. Building a regulatory environment for sustainable access and use of digital data and AI and establishing ethical frameworks and a common set of principles regarding data privacy and protection is fundamental.

UN Global Pulse worked extensively over the past years to advance Data Privacy and Digital Ethics, leading the privacy policy work in the UN system.

From Data Privacy and Protection

In 2014, UN Global Pulse established a Data Privacy Advisory Group [11] (PAG), a forum to promote an ongoing dialogue on critical topics related to data protection and privacy comprised of experts from the public and private sector, academia and civil society. Contributions by members have led to the world's first set of practical instruments designed specifically to address the ethical, security and privacy challenges that arise in using big data for the public good.

These instruments are expected to represent a useful contribution to a future in which responsible use of big data is ensured by systems, standards and regulatory frameworks that define the conditions under which big data may be shared and guarantee accountability not only for misuse of data, but for failures to use it when it ought to have been used. There is an opportunity to use these instruments to inform national legal and policy frameworks to promote the use of big data for the public good while protecting people's privacy. The following are key instruments and policy guidelines developed under the leadership of UN Global Pulse that are used by the United Nations:

- UN Global Pulse Data Privacy and Data Protection Principles [12].
- Risk, Harms and Benefits Assessment Tool [13].
- United Nations Development Group Guidance Note on Data Privacy, Ethics and Data Protection: Big Data for the achievement of the 2030 Agenda adopted by all 32 UN agencies [14].
- Big Data for Development and Humanitarian Action: Towards Responsible Governance [15].

With a proven record of capacity building internally and externally, UN Global Pulse provided over the years technical support to the development of privacy legislation in different parts of the world. For example, UN Global Pulse provided expert input and training to the government of Uganda on the pending Data Privacy Bill that was ratified by the President on February 2019. At the same time, it is an observer to the EU Commission's Expert Group on Business to Government Data Sharing and will contribute to the EC report and frameworks for data sharing for social good.

Ripple Effect—Adoption of the First System-Wide UN Principles on Personal Data Protection and Privacy

UN Global Pulse is the Founder and Co-chair of the UN Data Privacy Policy Group (UN PPG), created to facilitate dialogue and knowledge sharing on data privacy and protection within the UN and the broader development and humanitarian space. The UN PPG formulated a the first set of system-wide UN Principles on Personal Data Protection and Privacy [16] (https://unsceb.org/privacy-principles) that set out a basic framework for the processing of personal data by, or on behalf of, the United Nations System Organizations in carrying out their mandated activities. The UN system officially adopted these principles on 11 October 2018. The Principles aim to: (i) harmonize standards for the protection of personal data across the UN System; (ii) facilitate the accountable processing of personal data; and (iii) ensure respect for the human rights and fundamental freedoms of individuals, in particular the right to privacy.

To Ethics and AI

Algorithm-based systems powered by big data and AI increasingly learn from and autonomously interact not only with their environments, but also one another, leading to patterns of behavior that cannot always be predicted in advance, nor be explained after they have occurred. This sudden increase in self-organization of intelligent systems, combined with a trend towards use of proprietary algorithm-based decision-making tools in areas such as loan approval, job qualification, and university admissions, has far-reaching consequences for individuals, businesses, governments and society. It also creates great difficulties in implementing principles of transparency of data and algorithms and raises a series of challenging ethical considerations such as the following: Should AI be able to make life-and-death decisions? For example, should AI alone be permitted to decide how autonomous drones or robots can save lives or choose whom to save in search and rescue operations? What data may be used in making such a decision? Where does the liability rest for harm caused by AI? How can we avoid the biases in decision-making by AI, perpetuating inequalities and discrimination in moments of emergency response? How can we ensure that a world of increasingly preemptive computing remains human-centered, protecting human identity, values and dignity? How do we ensure that the right to benefit from an opportunity is protected? [17].

To respond to these questions, UN Global Pulse embarked in a new area of work to support AI policies to incorporate ethics and human rights-based approach.

References

1. The tool was featured on Astro Pi Challenge/Digital Planet of BBC (https://www.bbc.co.uk/programmes/p04h8s3k) and on NPR News (https://www.npr.org/sections/alltechconsidered/2016/10/31/500072478/turn-on-tune-in-transcribe-u-n-develops-radio-listening-tool?t=1617741978657).
2. Pulse Lab Kampala, Using machine learning to analyse radio content in Uganda (UN Global Pulse, 2017), https://www.unglobalpulse.org/document/using-machine-learning-to-analyse-radio-content-in-uganda/.
3. D4R, Towards an Understanding of Refugee Segregation, Isolation, Homophily and Ultimately Integration in Turkey Using Call Detail Records, https://d4r-turktelekom.unglobalpulse.net/.
4. Pulse Lab Jakarta, Vampire, http://vampire.pulselabjakarta.org/.
5. P. Hidalgo-Sanchis, How Big Data can strengthen official statistics in Africa: A view from Pulse Lab Kampala (UN Global Pulse, 2016), https://www.unglobalpulse.org/2016/03/how-big-data-can-strengthen-official-statistics-in-africa-a-view-from-pulse-lab-kampala/.
6. I. Madamombe, Community radio: a voice for the poor, in Africa Renewal 19, 2 (2005), https://doi.org/10.18356/614eab52-en.

7. UN Global Pulse, Zimbabwe Biodiversity, Climate Change and Resilience, https://biodivers ity.unglobalpulse.net//zimbabwe/.
8. Global Forest Watch, http://globalforestwatch.org/.
9. Protected Planet, http://protectedplanet.net/.
10. UN Biodiversity Lab, https://www.unbiodiversitylab.org/.
11. UN Global Pulse, Expert Group on Governance of Data and AI, https://www.unglobalpulse. org/data-privacy-advisory-group.
12. UN Global Pulse, Data Privacy, Ethics and Protection Principles, https://www.unglobalpulse. org/policy/privacy-and-data-protection-principles/.
13. UN Global Pulse, Risks, Harms and Benefits Assessment, https://www.unglobalpulse.org/pol icy/risk-assessment/.
14. United Nations Sustainable Development Group, Data Privacy, Ethics and Protection: Guidance Note on Big Data for Achievement of the 2030 Agenda, https://unsdg.un.org/resources/data-privacy-ethics-and-protection-guidance-note-big-data-achievement-2030-agenda.
15. Global Pulse Privacy Advisory Group, "Big Data for Development and Humanitarian Action: Towards Responsible Governance" Report (UN Global Pulse, 2016). https://www.unglob alpulse.org/2016/12/big-data-for-development-and-humanitarian-action-towards-responsible-governance-report/.
16. UN Personal Data Protection and Privacy Principles. https://www.unsceb.org/CEBPublicFiles/ UN-Principles-on-Personal-Data-Protection-Privacy-2018.pdf.
17. Copyright by UN Global Pulse.

Building the Field of Data for Good

Claudia Juech

Introduction

Previous chapters in this book have described compelling examples of how the use of data can be a powerful tool in making progress on some of humanity's biggest challenges. This article aims to outline how nonprofits, funders, governments, the private sector, citizens and consumers can collaborate to create the conditions for a sustained use of Data for Good.

Some Definitions

Data refers to the category of data narrowingly labeled "Big Data", which is characterized by its

- Volume—the amount of data,
- Velocity—the speed with which new data is generated or refreshed,
- Variety—the different types of data available
- validity and veracity—the reliability of the data source and how suitable the data is for its intended use.

For the purposes of this chapter, 'use' of data signifies the sustained ability to gather, analyze and act on data insights in an ethically responsible way and within legal frameworks. Lastly, 'Data for Good' focuses specifically on the not-for-profit

The author would like to thank Jade Wong for her research assistance, Louise Lopez for language editing and proofreading, and Stefaan Verhulst, Ananthan Srinivasan and Hazem Mahmoud for their input to earlier drafts.

C. Juech (✉)
Vice President Data and Society, Patrick J. McGovern Foundation, Santa Clara, USA
e-mail: claudia.juech@mcgovern.org

© The Author(s), under exclusive license to Springer Nature Switzerland AG 2021
M. Lapucci and C. Cattuto (eds.), *Data Science for Social Good*,
SpringerBriefs in Complexity, https://doi.org/10.1007/978-3-030-78985-5_4

application of data regardless of the size or areas of focus of the organization, e.g. by civil society organizations towards achieving the United Nations Sustainable Development Goals [1] or a local nonprofit working to protect the environment. It does not include data use by nonprofit universities for academic purposes.

Why Data for Good Matters

Our ability to utilize (big) data—ever larger quantities of data from diverse sources, such as satellite imagery, mobile phone data, social media and census information—in more sophisticated ways and faster than ever before; has opened up tremendous opportunities to:

- More fully understand complex social and environmental challenges and therefore be better able to address what's causing them.
- Use limited resources more effectively if the data can be used to identify and target subpopulations who are most at risk for diseases, persecution, abuse, or malnutrition or most affected by events such as a disaster or famine.
- Produce solid evidence more quickly verifying what works and what doesn't work leading to a more efficient delivery of services.
- Detect patterns that weren't visible using earlier methods to generate truly new insights and ways to solve problems.
- Use data to create accountability and align disparate actors around a common goal.

At the same time we need to be aware of and work on mitigating the risks that the use of data entails. The violation of privacy rights, for example, is not only a legal matter but threatens the freedom and lives of individuals. There is also the misperception of data as being neutral. The choice of aspects that are measured, collected and analyzed are driven by a purpose and invariably not unbiased. While derived from applied mathematics, such as statistics, data is about human beings and their behaviors [2].

Organizational Challenges Towards a Sustained Use of Data for Good

While there have been many projects in recent years that have demonstrated how nonprofits use of data can help them achieve their missions, the vast majority of these projects ended after the initial pilot phase. There are many factors to consider when trying to determine why organizations are finding it hard to move beyond the experimental stage. One key factor is the fundamental difference between extending the lifespan of an experiment and moving a data pilot successfully into production.

The use of data without parallel change management activities, including adjusting business processes, will likely only yield efficiency gains and miss broader impact across a nonprofit's mission. The sustained use of data within an organization requires deep operational, cultural and strategic changes as illustrated below.

	From pilot	To production
Organizational structure	*Separate reporting and budget lines for Data for Good projects set up*	*Data for Good elements incorporated into each major program or provided by a cross-functional service unit*

Data for Good pilots are often initially conceived by innovation units or IT departments and implemented as experiments outside of regular program lines. This structure provides an agile environment for testing new approaches. But nonprofit organizations and funders need to ensure that programmatic champions, decision criteria and processes are in place from the beginning to transition a successfully piloted approach into a standard element of programmatic work.

Staffing	*Temporary consultant, seconded employee or volunteer*	*Permanent employee(s) with appropriate expertise and experience*

Pilots inherently lend themselves to be staffed temporarily thereby making it more challenging for nonprofits to build their internal data capacities and knowledge base over time. Setting up an in-house data unit to run pilots and establish data routines across the organization could be one way of creating the needed internal data capacity. Deep understanding of an organization's data, its strengths and limitations is essential to identify where Data for Good workflows can contribute the most value. A central data steward function [3] can help to determine that value proposition, guide the mapping of data assets and ensure the responsible use of data. A data steward can also play a critical role in engaging with internal data holders and users to maximise use and benefits.

Technical Infrastructure	*Little to no capacity to collect, store, and analyze data*	*Data management practices and technology support current and future use cases*

The sustained use of data requires the development of a data strategy at the program and organizational levels. This entails the introduction of data management practices that allow for the storage and processing of growing amounts of data. These practices must also include the different data types, and the flexible execution of different use cases while ensuring compliance with legal requirements and ethical standards. Pilots run outside of a production environment are often not set up with future needs in mind and may only meet a few of these requirements.

Data collection and access	*Collecting data solely for operational or one-off purposes*	*Outlining specific analytical goals and a plan for collected data*

Many programs currently collect data, but may not be gathering data that can be further analyzed or used to draw new insights. Collecting data for analytical purposes—beyond the objective of supporting current operations or facilitating an ex-post evaluation—is important to ensure that the insights are relevant and can be used in real-life applications. Important steps guiding this data collection include determining the questions the data will help to answer, what data will be needed to do so, how to gain access, and *consistently* having access to the data. Successful data environments using external data will have put in place long-standing data sharing agreements whereas external data used in pilots is often only accessible for a short period of time. The acquisition of those data sets, at a cost or for free, requires sophistication in setting up data partnerships whose absence is often an important reason for not scaling or sustaining.

(continued)

(continued)

	From pilot	To production
Security and risk management	*No or minimal effort spent on creating additional measures to secure the data*	*Policies, procedures and software in place to guard data and analytics processes*

Most organizations do not take security into special consideration during the pilot phase and that lack of consideration can carry over to the production stage.

While dealing with common issues such as intrusion detection and prevention; big data security needs to pay extra attention to areas like encryption and user access control. For example, encryption tools need to secure data in-transit and at-rest, and be able to do so across massive data volumes, operating on many different types of data, both user and machine generated, as well as different analytics toolsets. Backups, including disaster recovery planning, and backup policies are another critical component often overlooked in pilots and that omission is carried over to production environments.

More generally, organizations need to adopt new or expanded policies on how to manage data responsibly. This practice in needed not only to ensure compliance with laws and regulations but also to uphold the ethics of data management and treat the humans behind the data with respect and dignity. Oxfam has given this much thought and developed a responsible data management (RDM) training pack available online [4] that introduces the principles of RDM. The training pack covers options for the planning processes and how to handle unexpected issues that arise in different contexts.

Data quality and analysis	*Demonstrating techniques to analyze data*	*Effectively iterating on approaches and drawing actionable insights from consistent and reliable data*

Pilot projects commonly address data quality issues by making a one-time effort to clean data. In contrast, a production environment requires implementing an automated data quality process to ensure ongoing data consistency and reliability. However, being able to produce useful insights goes beyond the data itself. It requires nuanced understanding of the data in context and the how and why of advanced techniques including the tools to be able to minimize biases, gaps, and imperfect data. Creating an algorithm that can support real-life decision making takes many iterations and will often exceed the narrower scope of a pilot project. This is not to say that all risks can be mitigated when working with data.

Funding	*Seed funding to prototype the use of Data for Good and distribute lessons learned*	*Funding available to support data engineering and analytics as a core nonprofit function*

This point touches on the broader challenge of nonprofits to move beyond bounded project support and obtain non-earmarked funding for core functions. Making the shift from investing in pilots to supporting the sustained use of Data for Good requires a clear articulation of the value add to the achievement of programmatic goals. It also requires considering possible savings through efficiency gains in addition to increasing the effectiveness of program strategies. In the medium-term those savings could offset the expenses needed to establish the new data science, engineering functions and infrastructure.

Most Data for Good projects face all of the challenges described above. Overcoming these at the individual organization level requires, first and foremost, the willingness of senior leadership to build and support a data culture where leadership

- Prioritizes and invests in data management and analysis

- Frequently makes the connection between data insights and the decisions they led to
- Furthers the data literacy of all staff
- Supports and encourages employees to access and derive insight from the organization's data [5].

As illustrated by the "From Pilot to Production" overview, the sustained use of data within an organization requires deep changes to enable staff to use data in an improved way to solve problems and make decisions. These changes affect workflows and operational processes. In addition, there are considerable cultural barriers to overcome, including

- Legitimate concerns that an increased and advanced use of data might expose the organization to reputational risks
- Hesitation to open up the organization to more transparency and scrutiny
- The practice of assessing program performance more frequently and quantitatively, while focusing on decision making that is more than ever informed by data and information.

Ultimately, the process requires leadership and vision at the executive level and across an organization to invest in the current as well as future value of this fundamental change in the face of the pressing urgency of today's problems which are compounded by constrained resources and a preference for project-based funding.

Why Build the Field of Data for Good?

In the last section I described the challenges to implement Data for Good approaches from a nonprofit's internal organizational perspective. However, that view on its own may be too narrow as it does not take into account the contextual factors that prevent nonprofits from building the sustained capacity for an advanced use of data. In this section, I zoom out to look at the enabling environment of Data for Good and the field as a whole.

"A field is a community of organizations and individuals working together to solve a common set of problems, develop a common body of theory and knowledge, and advance and apply common practices [6]." To quote Jim Canales on the general importance of field building—"sustained change comes from having a strong"—and I want to add diverse—"community coalesced around an issue [7]." Despite the amount of media coverage and conferences focused on the Data for Good topic, the field itself is barely a decade old. Theory and practice of all of its components are still in the early stages and continue to evolve rapidly as illustrated by the following facts:

- Talent & skills

  - The job title of a data scientist was coined only recently in 2008 by D.J. Patil and Jeff Hammerbacher, one of the co-founders of Cloudera Inc [8].
  - While the number of data science degree programs continues to grow exponentially, many schools did not see their first classes graduate until 2018.

- Regulation

  - The EU General Data Protection Regulation (GDPR) [9], the most important change in data privacy regulation in 20 years, was just implemented in 2018.

- Technology

  - Hadoop, the key open-source software framework for storing and processing large volumes of any kind of data, was in production use in 2006 but wasn't commercially distributed until 2009.
  - In addition to the explosion in mobile phone penetration, satellites and sensors (two other technologies that have been essential to the growth of data) have become much more affordable due to recent technological advances and new business models. Large and expensive satellites have been replaced by much smaller versions that can fly as secondary payloads, significantly lowering costs. And, the average cost of the Internet of Things (IoT) [10] sensors dropped by more than half, from $1.30 to $0.60 between 2004 and 2014, and prices are expected to shrink another 37% to $0.38 by 2020 [11].

The fact that the field of Data for Good is still in its early stages compounds the specific challenges nonprofits face. They cannot rely on a strong enabling environment to boost their organizational ability to include an expanded use of data in their programming.

What's Needed to Build the Field of Data for Good?

Funders, governments, academia, the private sector, citizens and consumers, individually and jointly, have a role to play to overcome these challenges and advance the field of Data for Good. Based on the earlier definition of what constitutes a field this involves (a) solving a common set of problems, (b) developing a common body of theory and knowledge and (c) advancing and applying common practices. The following suggestions concentrate on closing key gaps in these three areas and putting conditions in place that enable nonprofits to move toward a more sustained and advanced use of data. The list is not meant to be comprehensive.

Tackling a common set of problems

From a funder's perspective, while grantees are normally selected for their expertise and experience, in the field of Data for Good, most nonprofit organizations are still lacking both. As a consequence many struggle to determine where and how data can best help to make progress on programmatic challenges.

Research such as GovLab's upcoming list of "100 problems that can be solved by using data" [12] can provide useful guidance to identify the most promising data use cases. More generally, it appears critical for any field building ambitions to move beyond reporting on individual examples and focus more on broad efforts to produce generalizable insights across types of data, types of use cases and data analytical approaches.

In addition to packaging these insights in easily digestible ways for various audiences, e.g. executives, program staff and IT, there is a need for learning networks and communities of practice that facilitate exchanges which go deeper than common conferences. Organizations such as Nethope and its Center for the Digital Nonprofit or Techsoup are some of the platforms that could provide a safe space for those interactions helping nonprofits and other actors to align around a common set of challenges.

At the systems level, most funding seems to focus on tackling challenges related to data governance, data standards and the ethical use of data. While these issues are critically important, addressing them alone will not unleash the potential of data for impact nor will they by themselves lead to a more sustained use of data by nonprofits. In looking at the full data value chain, from collection to analysis, dissemination, and the final impact of data; more attention is needed on the uptake and linkages to decision making processes [13]. Also, to advance the field of Data for Good, the aforementioned data governance frameworks and ethical principles will need to be translated into practical guidance for nonprofits. This is necessary to help them develop the processes and mechanisms to confidently weigh the risks and benefits of more advanced data use. For example, many nonprofits have stopped or slowed down new data activities in response to fundamentally useful regulations such as GDPR because they are unclear about their risk exposure.

Building a common body of theory and knowledge

A mature field is characterized by a systematic body of knowledge, grounded in empirical evidence, which can be used for explanatory and predictive purposes and be verified or contradicted [14]. The field of Data for Good still needs to build a common body of knowledge. It needs to begin more rigorously evaluating the impact of data projects. Further evidence is needed to understand the value the use of data contributes as well as why and where it fails to do so.

Building stronger connections between academia and practice is a second area that, if it received more attention, could further advance the field of Data for Good. Academic researchers, especially in the Global North, have often privileged access to data and more advanced data science capacity. However, despite good intentions, research agendas are more guided by academic requirements than real world needs. The result is the lack of a robust system of producing new knowledge that is useful and relevant. If academic research and needed insights for nonprofit activities were better aligned with the dissemination of findings targeting practitioners, this would have strong potential to drive further impact. One of the examples that seems to be headed in the right direction is the Rights Lab program at the University of Nottingham, a large-scale research platform for ending slavery. Their Data Programme's Slavery from Space initiative uses geospatial observation, including machine-learning techniques, to map and measure slavery to uncover sites and industries with high levels of slavery. From the beginning, the interdisciplinary research teams work closely with nonprofits, governments and the private sector to respond to a common set of challenges such as, "how many slaves exist in the world and where are they?" and "what are the means for ending slavery [15]?" By Right Lab choosing to explicitly make practical application its main purpose, it provides nonprofits with information they can use to inform their interventions on the ground.

Another obvious gap hampering the creation of a common body of theory and knowledge is the lack of data science and data engineering talent. This is true in general but especially in low-income countries. Solving this challenge is no easy feat and not unique to the field of data. To create the diverse talent pool that is needed to meet the global demand entails closing *fundamental* gaps in different levels of education and tackling deficits in math and tech literacy. Improving access to affordable, good quality education would benefit many other sectors such as health, economic development and science beyond the field of Data for Good. This is a large, multi-generational task. At the same time there are more activities that can be done now.

Specifically include a strong capacity building component in projects from the beginning

Good examples are Care International's Bihar Technical Support Program or Terre des hommes' IeDa [16] project with the Ministry of Health in Burkina Faso where a great deal of emphasis has been placed on enabling local actors to manage and advance the approaches on their own at the end of the engagement.

Increase the availability of Data Science job training

Developing and implementing data science solutions requires practical experience. Year-long courses that include practical components and internships strike a balance between theory and practice and have real potential for reducing the talent gap in the interim. These data science schools are ideally set up in collaboration with universities so that they do not sideline existing education systems.

Think of functions and roles

Data science roles and responsibilities are not restricted to data scientists. In fact, the job roles and tasks are diverse in nature. The skills required for each job title, data engineer, data steward, data analyst, etc., vary significantly. However, most nonprofits are not taking a systematic approach to building their data team. Upskilling and retraining in house talent should be an option more widely utilized by non-profits. Not only does it expand the talent pool it is also a means of increasing job satisfaction and retention rates as demonstrated by Candid, the former Foundation Center in New York. They filled most of the roles on its 50+ person data team with existing staff members.

Advancing and applying common practices

As described earlier in the "From Pilot to Production" overview, the field of Data for Good is still predominantly characterized by fragmented one-off activities. This section aims to highlight where the development of common practices could accelerate the field building process.

- In the private sector

Pilot projects have shown that companies in sectors such as telecommunications, social media, satellite imagery and finance have proprietary data assets that can be utilized for greater societal benefit. However, there is no unified approach to making insights from those data assets available to nonprofits. Establishing common practices would help to reduce the transactional costs per instance and increase the transparency of how external requests for data are handled. GovLab's efforts to build a network of corporate data stewards is one step in the right direction. The network includes individuals or teams tasked with identifying, structuring, guiding and evaluating collaborative data opportunities [17]. More work is still needed to design and test business models that make data (insights) available to nonprofits on an ongoing basis. For example, could consumers be asked to give consent for their data to be shared for specific nonprofit purposes, such as health data to support cancer research? In other areas, voluntary industry-wide agreements might not reach far enough. The sharing of data insights should not predominantly be informed by business objectives and the promotional value of data collaborations. It should be driven, first and foremost, by impact considerations and available on the same terms to all nonprofits.

- In the nonprofit sector

Since its inception in September 2017 through March of 2019, the team at the Cloudera Foundation, now the Patrick J. McGovern Foundation, spoke with more than 300 nonprofits about their data needs and challenges. These nonprofits represented a wide range of disciplines and fields indicating that the advanced use of data is becoming more and more relevant in areas as different as human trafficking, health care, wildlife tracking, climate change, poverty alleviation, and disaster response. At the same time, as described earlier, all of these nonprofits struggle to build internal data engineering and data science capacity. This will remain a challenge that most organizations will not be able to address on their own in the coming years.

Data-focused collaborations between nonprofits could present a potential solution. They would enable organizations to:

- **Improve access to technical resources and expertise**: While the growing availability of cloud-based services will make it easier to enter this field, nonprofits could mitigate some of the challenges in this area by establishing shared 'data centers' that provide data engineering and data science capacity to a group of organizations allowing a reduction in costs while optimizing the use of expensive resources.

– **Improve the likelihood of generating useful insights**: Many nonprofits' data sets are, on their own, not suited for predictive or prescriptive analyses because of their small size and lack of variety. Data collaboratives could potentially fill that gap by augmenting and enriching the data each organization has access to individually without raising concerns about a for-profit use of data.

Collaborations between nonprofits are rare, for many reasons, not least of which because organizations compete for the same sources of funding. Funders need to explore ways to stimulate data collaborations between nonprofits and look beyond supporting first movers if they want to contribute to building a field where data is commonly used to improve impact.

Advancing and applying common practices is equally relevant *across* sectors. Most nonprofits are eager to learn about corporate use cases and adopt practices. Corporate volunteers can play an important role in dispersing this knowledge. However, to avoid the pitfalls of pilots and one-offs, the involvement of corporate volunteers needs to be carefully planned. It should be part of a nonprofit's longer-term data strategy and roadmap. Matchmaking organizations such as DataKind understand the benefits and challenges of the volunteer approach. They have made significant progress in finding ways to embed pro bono data scientists in nonprofits for longer periods of time to increase the likelihood of success.

• In the public sector

Whether data is being used commercially or for the social good, governments play a critical role. Cities, states and national governments are not only the creators of the legal and regulatory environment guiding critical issues such as data ownership, privacy and security. They are also key data owners, providers and data users. This section can only raise a few core aspects of how cities, states and national governments can advance the application of common practices in the field of Data for Good.

- **Be inclusive**: Incorporate the voice of vulnerable populations in the design of data policies and ethical guidelines so that these populations can harness the benefits of increased data use while keeping risks at a minimum.
- **Open Data**: Make as much data produced or collected by government entities freely available so that it can be easily consulted and re-used. To generate intelligence, nonprofits—beyond using their own data—increasingly rely on new datasets provided by cities, states, federal governments, and multilateral organizations, etc. While these developments are a good starting point, it should not be overlooked that citizens and civil society organizations in many geographies still face significant barriers to utilize open data such as limited or very costly internet access.
- **Emphasize data quality**: As key collectors, providers and users of data that is widely used to inform resource allocations, investments and the provision

of goods and services governments at all levels have a special responsibility to focus on collecting accurate and representative information. When there are potential errors or limitations in the data, e.g. under- or overrepresentation of certain populations, these factors should be clearly noted to help prevent existing biases from being replicated and amplified by data use.

The importance of the public sector as the originator of data sets and as a user cannot be overemphasized. The more governments, including cities, establish guidelines, collect data and model user behavior with the differentiated needs of all citizens in mind, the more they are helping to shape good common practices for the field of Data for Good.

- *In the funding community*

The nonprofit sector will need core support and infrastructure funding if the goal is to establish the use of data as a core competency. But to help advance the application of common practices—as part of building the field of Data for Good—foundations and other funders can also draw on some of their traditional strengths:

– **Be a trusted convener across sectors**: Provide safe spaces to exchange different perspectives and develop a common understanding around different areas of work such as data privacy, policy and technology.
– **Support the generation and diffusion of practical knowledge**: Support impact evaluations of data activities to help generate generalizable evidence; model behavior and share what has worked and what has not.
– **Foster collaboration**: Build networks within and across the nonprofit sector; support the development of open standards to assist with the collaboration.

Collaboration within philanthropy itself and across the broader spectrum of donors is equally important to help build the conditions for systemic change. This will ensure organizations and individuals are equipped to act when the use of data yields new insights. This means, for example, having sufficient shelter beds available to accommodate additional demand if an organization combating human trafficking is successful in using advanced data approaches to identify victims.

Conclusion and a Look Ahead

It is time to move beyond thinking of Data for Good as building the capacity of individual organizations and place equal emphasis on creating an enabling environment for an advanced and sustained use of data. This includes establishing clear boundaries to strike the right balance between realizing potential and protecting everyone—especially vulnerable populations—from harm and exploitation. These rules of engagement—whether they are laws, regulations or voluntary principles—need to be made with the future in mind as the sector and possible use cases are quickly evolving.

From a funder's perspective building the field of Data for Good requires many different competencies and interventions, making a strong case for increased transparency and collaboration among funders. "Some funders focus on research, some on building and testing models, and some on other things. If we know more about how those pieces fit together and where pieces are missing, we can create a value chain where everyone knows where they fit in, and where more time, attention, and other resources need to be invested [18]." If we apply this approach to building the field of Data for Good we have a strong chance of permanently adding a potentially transformative tool for achieving positive impact.

References

1. UN General Assembly, Transforming our world: the 2030 Agenda for Sustainable Development, A/RES/70 (2015).
2. M. Ecowo, Why Numbers can be Neutral but Data Can't (New America, 2016). https://www.newamerica.org/education-policy/edcentral/numbers-can-neutral-data-cant/.
3. Medium, Data Stewards, https://medium.com/data-stewards-network.
4. R. Hastie, A. O'Donnell, Responsible Data Management training pack (Oxfam Guidelines and Toolkits, 2017). https://policy-practice.oxfam.org.uk/publications/responsible-data-management-training-pack-620235.
5. C. D'Ignazio, R. Bhargava, You Don't Need a Data Scientist, You Need a Data Culture (Datatherapy, 2017). https://datatherapy.org/2017/12/06/building-a-data-culture/.
6. K. O'Neil, How Do You Build a Field? (The Rockefeller Foundation, 2015). https://www.rockefellerfoundation.org/blog/how-do-you-build-a-field-lessons-from-public-health/.
7. J. Canales, From the President: Building A Field (The James Irvine Foundation: Our Blog, 2013). https://www.irvine.org/blog/from-the-president-building-a-field.
8. T. H. Davenport, D. J. Patil, Data Scientist: The Sexiest Job of the 21st Century (Harvard Business Review, 2012). https://hbr.org/2012/10/data-scientist-the-sexiest-job-of-the-21st-century.
9. More on EU General Data Protection Regulation: https://eur-lex.europa.eu/eli/reg/2016/679/oj.
10. CoWorkr & Haworth AP White Paper, How Embedded Sensors will Transform Workplace Performance, Employee Engagement, and Facility Management (CoWorkr, 2017). https://medium.com/coworkr/how-embedded-sensors-will-transform-workplace-performance-employee-engagement-and-facility-2a3fb9ae1142.
11. M. Baylor, Planet Labs targets a search engine of the world (NASASpaceFlight, 2018). https://www.nasaspaceflight.com/2018/01/planet-labs-targets-search-engine-world/.
12. The 100 questions, https://the100questions.org/.
13. Open Data Watch, The Data Value Chain: Moving from Production to Impact (Data2X, 2018). https://data2x.org/resource-center/the-data-value-chain-moving-from-production-to-impact/.
14. M. N. K. Saunders, D. E. Gray, P. Tosey, E. Sadler-Smith in *A Guide to Professional Doctorates in Business and Management* ed. by L. Anderson, J. Gold, J. Stewart, R. Thorpe (Sage, 2015), pp. 35–56.
15. The Rights Lab, https://www.nottingham.ac.uk/research/beacons-of-excellence/rights-lab/index.aspx.
16. IeDa: Improving primary health care through eHealth, https://ieda-project.org/.
17. S. Verhulst, The Three Goals and Five Functions of Data Stewards. Data Stewards: a new Role and Responsibility for an AI and Data Age (Medium, 2018). https://medium.com/data-stewards-network/the-three-goals-and-five-functions-of-data-stewards-60242449f378.

18. R. Guthrie, E. Cooperström, The 450 Million Farmer Opportunity: A Conversation with the Skoll Foundation Team (The Skoll Foundation. Inside the Issue, 2015). http://skoll.org/2015/10/11/the-450-million-farmer-opportunity-a-conversation-with-the-skoll-foundation-team/.

When Philanthropy Meets Data Science: A Framework for Governance to Achieve Data-Driven Decision-Making for Public Good

Nuria Oliver

Introduction

Philanthropy means *love for humanity*. Today, the term refers to the economic and social sector consisting of private initiatives (in the form of foundations, donations, non-profit organizations and so forth) who devote resources for the public good, that is, to improve the welfare of others in an altruistic way.

Looking at the figures, one could think that the philanthropy sector is living a golden period. According to the 2018 UBS Global Philanthropy Report [1], there are 260,000 foundations in 38 countries worldwide with total assets of about 1.5 trillion USD and an average annual spending that exceeds 180 billion USD. Education is the highest priority, representing 35% of the total investment and being the focus of almost 30,000 foundations. Other topics of interest include social welfare (21%), health (20%) and culture and arts (18%). Interestingly, Latin American foundations have started a process of aligning their strategies to the achievement of the 17 Sustainable Development Goals (SDGs).

Looking at these figures, we could say that we live in a *world of philanthropy* and the impact that the philanthropy sector is having and could have on society is huge.

But we also live in a world of data, of Big Data, which we could use to support philanthropy by, e.g., informing our decision-making regarding philanthropy projects and enabling the measurement of the impact and progress of such projects.

In this chapter, I establish a connection between these two worlds with a focus on the key dimensions that we should address before we can fully realize the potential of leveraging Big Data for humanity. Note that parts of this chapter have appeared in previously published work by me and my coauthors [2–4].

Dr. Nuria Oliver (✉)
ELLIS Unit Alicante Foundation, Alicante, Spain
e-mail: nuria@ellisalicante.org

Data-Pop Alliance, New York, NY, USA

© The Author(s), under exclusive license to Springer Nature Switzerland AG 2021
M. Lapucci and C. Cattuto (eds.), *Data Science for Social Good*,
SpringerBriefs in Complexity, https://doi.org/10.1007/978-3-030-78985-5_5

Big Data for Social Good

Today we have unprecedented access to massive streams of human behavioural data, because of the exponential growth in the adoption of mobile devices and the increased digitalization of the physical world via sensors and connected objects (which has led to the Internet of Things revolution). This data, combined with significant advances in analytical capabilities (particularly in data-driven machine-learning methods) and access to affordable high-performance computing, is enabling companies, governments, and other public sector actors (including foundations and NGOs) to use data-driven machine learning-based algorithms to tackle complex policy problems [5].

Thus, decisions with both individual and collective impact that were previously taken by humans—often experts—are nowadays taken by data-driven artificial intelligence systems (i.e. algorithms), including decisions regarding the granting of credits and loans, the hiring of people, judicial judgements, policing, resource allocation, medical diagnoses and treatments, the admissions of students in universities, and the purchase/sale of shares in the stock market. Data-driven algorithms have the potential to improve our decision making in key public good areas which are of relevance to the philanthropy sector.

Why would we want to rely on algorithms to make such important decisions? Because as history has shown, human decisions are not perfect. They are subject to conflicts of interest, corruption, selfishness, greed, hunger, and cognitive biases, which have led to unfair and/or inefficient processes and outcomes [6]. Therefore, the interest in the use of algorithms can be interpreted as the result of a demand for greater objectivity in decision–to meaningful human contact—for example, in services operated exclusively by chatbots—and the right to not making and for a better understanding of our individual and collective behaviours and needs.

Data-driven algorithmic decision-making may indeed enhance the overall efficiency and public service delivery of philanthropies by helping to optimize bureaucratic processes, by providing real-time feedback, by predicting outcomes [7] and by helping to uncover and quantify existing problems and challenges in society. Some authors, like Parag Khanna in his book *Technocracy in America*, believe that a data-driven direct technocracy would be a superior alternative to today's representative democracy, because it could dynamically capture the specific needs of the people while avoiding the distortions of elected representatives and corrupt middlemen [8].

The potential for data-driven algorithmic decision-making to make a positive impact in the world is massive and certainly motivates my work on this area for over 10 years both in the private and public sectors [9–13].

Fortunately, in the past decade numerous efforts worldwide have emerged to explore this potential. Examples include the New Deal on Data led by Alex Pentland at the World Economic Forum, which is focused on consensus policies and initiatives to give citizens control over the possession, use and distribution of their personal data; NGOs such as the Partnership on AI, Data-Pop Alliance (where I am Chief Data Scientist) and Flowminder, which are focused on leveraging large-scale data and machine-learning techniques for social good in areas such as financial inclusion, public health and climate change/natural disasters; United Nations initiatives including the World Data Forum, the Global Partnership for Sustainable Development Data and Global Pulse; OPAL, a project led by Data-Pop Alliance with the goal of taking advantage of big data and artificial intelligence for social good while preserving the privacy of people, in a sustainable, scalable, stable and commercially viable way; private sector initiatives in telecommunication companies or banks; the GSMA Big Mobile Data for Social Good initiative, led by the GSMA and the United Nations Foundation, in which 20 mobile operators participate to contribute through the analysis of aggregate and anonymous mobile data to solve problems in the areas of public health and climate change/natural disasters; and the AI for Good Global Summit of the ITU, an international summit of United Nations for the dialogue on artificial intelligence, aimed at identifying the practical applications of AI for the improvement of the sustainability of the planet. The latter is managed by the International Telecommunications Union (ITU) as a specialised agency of United Nations for information and communication technologies.

Challenges

However, data-driven decision-making for social good is not exempt of limitations and challenges. If it were easy, we would already be using data to measure impact of projects, identify new opportunities, better allocate resources, optimize processes, and so forth. The challenges can be structured in four large categories described below. A more detailed description of existing challenges can be found in Letouzé, 2019 [14]:

Legal and regulatory challenges as most of the data that is valuable for public interest purposes has been collected for other purposes (a large portion of it for the provision of digital services). Hence, different uses of the data even if for social good might not be included or allowed by existing legislation or user consent. Moreover, if want to encourage and support the development of this new way of decision-making, we might need to update or design new regulatory frameworks that incentivize and enable the use of data for public interest goals. At the European Union level, the European Commission created in 2018 a high-level expert group on "Business to Government Data Sharing" that I am a member of. The report [15] prepared by this group was published in February of 2020 with key recommendations to policy makers to create the necessary ecosystem that would foster more data sharing from the private to the public sectors for public interest purposes.

Skills and capacities challenges given that data is just the starting point. The goal is to extract meaningful insights from the data via the use of machine learning and advanced analytics algorithms. Hence, teams with the right skills and capacities are necessary in the philanthropy sector who can transform data into knowledge. Unfortunately, there is a lack of professionals with expertise in data science, machine learning and related fields across all sectors and particularly in the philanthropy sector.

Technical challenges derived from the fact that this new approach of making decisions is not fool proof. There are challenges related to the actual access, storage, and processing of large-scale, dynamic, varied and constantly growing data in a secure and privacy preserving way. There are also algorithmic challenges, such as algorithmic bias and discrimination, unclear accountability, opacity, and limited robustness of state-of-the-art systems which can be hacked and fooled. In addition, other technical challenges emerge from the difficulties associated with the need to combine multiple datasets—ith different time scales, granularities, and levels of noise—to tackle real problems, the lack of existing ground truth, the difficulties to infer causality (but rather correlations) or the complexity of running experiments in the real world.

Ethical challenges also abound. When algorithmic decisions affect thousands or millions of people, important ethical dilemmas arise. For example, does this mean that automatic decisions are beyond our control? What level of security do these systems need to have to protect themselves from cyberattacks or malicious use? How can we guarantee that the decisions and/or actions based on the use of such algorithms do not have negative consequences for people? Who is responsible for these decisions? What will happen when an algorithm knows each one of us better than we know ourselves and can take advantage of that knowledge to manipulate our behaviour subliminally? Plato's words over 2,000 years ago are surprisingly relevant today: "*A good decision is based on knowledge, not on numbers*".

Beyond preserving human rights, we find in the existing literature numerous proposals of ethical principles and working dimensions that should be addressed to ensure that data-driven decision-making has a positive impact on society. I summarize them in the FATEN framework described below.

FATEN Algorithms: When Data Science Meets Philanthropy

In this section, I describe the key pillars that I believe we should implement before we can fully realize the immense potential of data-driven decision-making for philanthropy. These pillars are grouped in the acronym FATEN for fairness, autonomy/accountability/augmentation, trust/transparency, equity/education/beneficence and non-maleficence. Note that the discussion below is largely based on the articles presented in Oliver [16], and Lepri et al. [17]:

(1) *Fairness*

Non-discrimination and fairness should be at the core of any system that relies on data-driven machine learning-based algorithms to support human decision-making [18]. Unfortunately, as we have witnessed in recent years, algorithmic decisions may discriminate because of biases in the data, because of inappropriate use of an algorithm and/or because of the characteristics of the algorithms themselves. Algorithmic decisions might not only reproduce but also amplify existing patterns of discrimination in society, which are a result of pre-existing biases or human prejudices [19]. One of the most well-known examples algorithmic discrimination is illustrated by the ProPublica study of the COMPAS recidivism algorithm which was used by judges in the US to inform criminal sentencing decisions by predicting the probability of an individual to reoffend. ProPublica's investigative journalists found that the COMPAS algorithm was significantly more likely to assign a higher probability of reoffending to black defendants than to white defendants, despite having similar prediction accuracies between both groups [20]. Cathy O'Neil's book *Weapons of Math Destruction* describes several case studies where data-driven algorithmic decision-making has caused harm and risks to public accountability, particularly in the areas of education and criminal justice [21]. Moreover, individuals might be negatively impacted by data-driven algorithmic decision-making processes not because of their own actions, but because of the actions of others with whom they share certain characteristics.

From a technical perspective, algorithmic fairness and non-discrimination is a highly active research area. Beyond investing in developing technical solutions to identify, quantify and combat algorithmic discrimination, it is of paramount importance to bring together experts from relevant disciplines—such as economics, the law, ethics, computer science, political science, and philosophy, to design and empirically validate appropriate metrics of justice to the real-world task at hand. To complement the necessary empirical research, there is a need for developing theoretical models that would help the users of such algorithms ensure that decisions are made as fairly as possible and mitigate potential negative, unintended consequences.

Due to the transversal nature of data-driven algorithms and their potential application to many areas of consequential importance in people's lives (e.g. education, healthcare, finance, immigration and criminal justice), I would like to emphasize the importance of *cooperation.* A constructive exchange of resources and knowledge between the public, private, and social sectors should be encouraged and nurtured. Cooperation is important not only between different sectors but also between nations—given today's globalisation, as emphasised by the well-known Israeli historian and thinker, Harari [22]. A lack of effective cooperation has been identified as a weakness in the philanthropy sector. In fact, about half of the foundations existing today are small: they do not have any paid employee and their budgets are below 1 million USD. Hence, their impact and reach are limited: opportunities might be lost due to lack of scale and missed synergies with other foundations. Increased coordination and cooperation in the philanthropy sector could lead to larger impact. Opportunities for cooperation might become apparent through data analysis about the activities and characteristics of philanthropies world-wide.

(2) *Autonomy, Accountability, and Intelligence Augmentation*

Autonomy is a central value in Western ethics. The principle of autonomy states that every person should be able to decide their own thoughts and actions, thus ensuring free choice and freedom of thought and action. However, this principle of autonomy is questioned today in our daily interactions with technology. Computational models of our needs, traits, taste, desires, and behaviour can be built through the analysis of our own behavioural data—our digital footprints resulting from our interactions with both the physical and digital worlds. Moreover, such models may be exploited by service providers to subliminally influence our decisions and our behaviour. Hence, we should ensure that autonomous decision-making systems always preserve human autonomy and dignity. For this, the systems need to behave in accordance with accepted ethical principles of the society where they are deployed.

Preserving human autonomy in automated decision-making systems is an active area of research. As a result, many guidelines for ethical frameworks have been proposed at different levels (national, supra-national, such as the European Commission and professional associations, such as the ACM or the IEEE). Nonetheless, there is no single recognised method for embedding ethical principles into data-driven algorithmic decision-making processes. Moreover, all experts, data scientists and professionals working on the development of Artificial Intelligence systems that affect or interact with people (algorithms for decision-making, recommendation and personalisation systems, chatbots, and so on) should be asked to behave in accordance with a clear code of conduct and ethics defined by the organisations where they work.

We also need clarity regarding the *attribution of responsibility* for the consequences of the actions or decisions taken by autonomous systems, as it is done with most of the services and products that we use and deploy in our society. While transparency and audits are key factors to contribute to accountability, they are not enough to guarantee the clear accountability of autonomous systems. Thus, special provisions for accountability are necessary.

Finally, I would like to highlight the principle of *intelligence augmentation* instead of intelligence substitution or replacement. It is constructive to adopt a synergistic vision between algorithms and humans, a vision where computational systems are used to increase or complement human intelligence, not to replace it. For example, an automatic translation system can be considered a system to increase our intelligence, since it expands our abilities with the capacity to translate from one language to another; an algorithm to automatically detect breast cancer in mammograms augments the intelligence of radiologists and oncologists by providing better-than-human detection capabilities which humans can use to make more informed decisions regarding their diagnoses and prescribed treatments.

(3) *Trust and Transparency*

Our relationships with other humans or with institutions depend on trust. Similarly, technology relies on the trust of its users who enrich and even delegate their (our) lives to digital services. Unfortunately, people's trust in the technology sector has been declining, partly fuelled by public exposés of privacy violations and abusive behaviour, such as the Facebook/Cambridge Analytica and the Huawei scandals. The philanthropy sector also needs the existence of a trusted relationship with all their stakeholders, from funders and donors to governments and citizens.

Trust emerges when three conditions are met: (1) first, we need to demonstrate *competence* regarding the specific task that the trust is deposited onto; (2) secondly, competence needs to be sustained over time, leading to *reliability*. It is not enough to do things well once; and (3) finally, there needs to be *honesty* and *transparency* in the relationship for trust to flourish.

In this context, *transparency* is understood as the property of a computer system or model to be easily understood by non-experts. As previously mentioned, transparency contributes to accountability. Despite the importance of transparency, algorithmic decision-making systems are not necessarily always transparent [23, 24]. According to Burrell [25], algorithmic decision-making systems might be opaque (i.e., might lack transparency) for three reasons.

First, they might be **intentionally opaque** to protect the intellectual property of the owners or creators of the systems. Intentional opacity might be mitigated by regulation and legislation which could enforce opening the software systems. An example of such legislation is the European General Data Protection Regulation (GDPR), which has defined a right to an explanation regarding human and algorithmic decisions. However, powerful commercial and governmental interests might make it difficult to address this type of opacity.

The second type of opacity is known as **illiterate opacity** which is the result of a general lack of technical skills and knowledge in society. Even if one were to reveal how an algorithmic decision-making system works, the explanation might not be sufficiently clear for a non-expert to understand it. This type of opacity is alleviated by developing educational programs in digital skills and by enabling independent experts to advise those affected by data-driven algorithmic decision-making processes.

Finally, algorithmic decision-making systems might be **intrinsically opaque** due to the characteristics of many state-of-the-art machine-learning methods (such as deep learning models [26]). This type of opacity has been extensively studied in the machine-learning research community. It is also referred to as the problem of interpretability or explainability.

Beyond algorithmic transparency, we also need to provide transparency related to what human behavioural data is captured and analysed, and for what purposes— which is addressed in the GDPR at the European level—and related to when humans are interacting with artificial systems (e.g. personal assistants or chatbots) versus interacting with other humans.

(4) *Education, Beneficence and Equality*

I have briefly described the capacity and skills challenge that we face and that we would need to tackle before we can fully benefit from data-driven decision-making for philanthropy. The need to invest in education is urgent, large, and transversal: *education* at all levels.

First, the compulsory education curricula by adding Computational Thinking as a core subject starting in primary school, coupled with an emphasis on nurturing our creativity, critical thinking, and our social and emotional intelligences. Computational thinking includes five core areas of knowledge: algorithms, programming, data, networks, and hardware. Given that 35% of the activities in philanthropic institutions today are devoted to education, a focus on narrowing the digital skills gap and on bringing today's schools to the 21st century could be of interest and impact. I describe my views on this subject in the chapter entitled "Digital Erudites" in the book "Digital natives do not exist" (title translated from its original title in Spanish) [27].

Education is also needed for our policymakers and politicians, for professionals in the philanthropy sector and particularly for those whose jobs will be affected by the development of Artificial Intelligence and for citizens so they can make informed decisions about systems that are having and will have an impact on their lives.

Moreover, the use of data-driven algorithmic decisions should always aim to maximize their **beneficence**, that is, their positive impact on society with *veracity, sustainability,* and *diversity.* This principle of beneficence is fully aligned with the goals, mission, and vision of the philanthropy sector.

I would like to emphasize that not every technological development entails progress. What we should strive for and what we should focus on is progress. To me, progress means improving the quality of life of people—all people, not just a few–, of the rest of living beings on our planet, and of our planet itself. Progress in this sense is the *raison d'être* of philanthropy. Let us look in a bit more detail these three dimensions: veracity, sustainability, and diversity.

In the past few years, we have witnessed the rise of *deep fakes,* that is, synthetically generated content—text, photos, videos, or audio—using deep learning techniques which is indistinguishable from real content. Deep fakes contribute to the worrying trend of *fake news,* that is, false or misleading information that is presented and broadly shared using social media platforms as actual news. Fake news might define public opinion on important social issues—such as who should be the next president of a country or whether a country should remain a member of the EU—to favour the interests of a minority that generates and disseminates such content. Thus, guaranteeing the *veracity* of both the data used to train machine-learning algorithms and the content we consume is of paramount importance. In fact, an assessment of the veracity of the data should be performed before training any data-driven model to assist in decision-making, to ensure that the models are trained with data that reflects an underlying reality. We should not disregard that well-intentioned projects, including projects in the philanthropy sector, might lead to negative consequences in the communities where they are deployed if certain precautions are not taken, including ensuring the quality of the data which the decisions are going to be based upon.

The technology sector broadly speaking, and data-driven decision-making systems in particular consume significant amounts of energy, with a clear negative impact on the environment [28]. Today's pervasive deep-learning techniques require exceptionally large computing capabilities to analyse massive amounts of data with prohibitive energy costs, especially if we consider the deployment of such systems on a large scale. Thus, we need to ensure that the technological development is aligned with our responsibility to maintain and ideally improve the living conditions on our planet and to preserve our environment for the generations to come.

At the same time, we need data-driven machine-learning algorithms to address many of the most important environmental challenges that we face (e.g., climate change, scarcity of resources, sustainability etc.). Artificial Intelligence enables us to develop key technologies, such as more efficient renewable energy systems (e.g. smart grids, efficient solar and wind energy), sustainable transportation (e.g., autonomous electric cars, for example) and precision agriculture.

Data-driven machine learning-based decision-making algorithms might be applied in a variety of use cases, as previously described, from education to healthcare and immigration. Unfortunately, such diversity of applications is not paralleled by a diversity in the teams that create them. In fact, there is a worrying lack of diversity in the teams that develop such systems, which tend to be composed of homogeneous groups of male computer scientists. Moving forward, we should ensure that the teams that are responsible for creating the technology that is so universally used are diverse both from the perspective of their areas of expertise and their demographics (especially gender, given that women occupy less than 20% of technical positions in most technology companies).

Another area where there is a lack of diversity is in the algorithms themselves. For example, personalisation and recommendation algorithms frequently lack diversity in their results. They tend to cast their users into groups according to their past patterns of consumption, leading to *filter bubbles* [29] and *echo chambers* that reinforce our own views (and prejudices) and make us distrust viewpoints different from our own. This lack of diversity in the people that we interact with and/or the content that we are exposed to and consume is undesirable. It limits the role of recommendation systems as tools that enable us to *discover content*—e.g., books, music, news, movies or even friends—that differs from our own tastes and that could help us broaden our thinking, understand other viewpoints, and encourage open-mindedness.

Finally, we need to reflect about *equality*. The 4th Industrial Revolution that are live in might entail a decline in our solidarity and spirit of equality, which would have a negative impact on the philanthropy sector. There is no doubt that the development, growth and wide access to the Internet and the World Wide Web have been instrumental to democratize the access to information worldwide. However, the principles of universal access to knowledge using technology are questioned today partly due to the dominance of a handful of technology giants in the US (Alphabet/Google, Amazon, Apple, Facebook, Microsoft) and in China (Tencent, Alibaba, Baidu). This situation of dominance by an oligopoly of companies has been coined as a *winner takes all* phenomenon. Together, these technology companies have a market value of more than 5 trillion USD. Their market shares in the US are unprecedented: more than 90% in Internet search (Google), more than 70% in social networking (Facebook) and about 50% in e-commerce (Amazon).

The XXI century is witnessing a polarization in the accumulation of wealth, which might have had a positive impact on philanthropy as many of the most active and most impactful philanthropic initiatives are funded by the multi-billionaires in the world. According to a study by Credit Suisse [30], the 1% richest in the planet owns 45% of the world's wealth and the richest 10% of people own 82% of the global wealth. This concentration of wealth in the hands of very few has been attributed, at least partially, to the 4th Industrial Revolution and the technological developments that have led to it.

In our history, the sources of wealth have evolved over time. During the Agrarian Revolution in the Neolithic period and for thousands of years afterwards, the ownership of the land entailed wealth. With the First Industrial Revolution in the XVIII and XIX centuries in the US and Europe, the ownership of factories and machines—rather than land—led to wealth. Today, one could argue that data—and more importantly the ability to derive insights and monetize such data—is the most valuable asset, leading to what is known as the data economy and to massive amounts of wealth for those who hold the data and are able to monetize it.

Let us not forget that from the five most populated *countries* in the world (Facebook, WhatsApp, China, India and Instagram), three are digital countries owned by Facebook. Countries with less than 20 years of existence, a global reach of billions of people, which are governed by a non-democratically elected president. As a result of this technology dominance by a few, a large percentage of today's data—and particularly human behavioural data, i.e., data about each of us—is **privately held**. Data that is constantly captured, analysed, and leveraged by these technology giants who not only know our needs, movements, interests, tastes and social relationships, but also our happiness or educational levels, our sexual or political orientations and even our levels of mental health. Data that could be of special value and relevance to the philanthropy sector.

Hence, to maximize the positive impact on society of the technologies that are at the core of the Fourth Industrial Revolution—and particularly of data-driven Artificial Intelligence techniques—we should develop new models of ownership, management, sharing, monetization, usage, and regulation of data. Europe's GDPR and its High-Level Expert Group on "Business to Government Data Sharing" are examples in this direction. However, the practical application of many of the proposed laws, principles and recommendations has proven to be complex, partly due to the difficulties derived from the fact that data is an intangible, varied, dynamic, distributed asset which is replicable infinite times at practically zero cost.

This implication is particularly relevant to philanthropy as a tool to bring back equality in today's increasingly unequal world. Today we can use large-scale, human-behavioural data to automatically infer and measure inequality, to predict socio-economic status, well-being and to promote financial inclusion, see for example San Pedro et al. [31], Hillebrand [32], and Soto [33].

And finally, we have the principle of non-maleficence.

(5) *Non-maleficence*

In addition to maximizing the positive impact, we should strive to minimize the negative impact that the deployment and use of data-driven decision-making algorithms might have on society, i.e., its maleficence. There are seven elements of the non-maleficence principle that I would like to briefly discuss: prudence, security and reliability, robustness and reproducibility and data protection and privacy. All of them are relevant in the context of data-driven philanthropy projects. Note that the European Commission published in February of 2020 a white paper on Artificial Intelligence [34] that outlines possible regulation of AI systems regarding a few of these important dimensions.

The principle of *prudence* emphasises the importance of considering different options in the initial phases of the design of any system to maximise its positive impact and minimise the potential risks and negative consequences derived from its application. Several requirements need to be met by the professionals that develop and deploy data-driven machine learning-based algorithms for decision-making, such as ensuring that high-quality data is being used to train the systems, the consideration of different working hypotheses that include a variety of perspectives, and the participation of multi-disciplinary experts to analyse and properly interpret the models and their results in context.

The vast majority, if not all, of the systems, goods, and products we consume—such as vehicles, clothing, toys, food, household appliances, medicines, medical devices, and industrial machinery, need to comply with strict quality, reliability and safety controls and regulations to minimise their potential negative impact on their users and/or on society. Data-driven algorithmic decision-making systems should also be expected to be subject to similar processes. Hence, it might make sense to create a European-level authority to certify the security, safety, and reliability of AI-based systems before they are widely deployed and commercialized in our society. Moreover, autonomous decision-making systems should provide guarantees regarding the integrity and safety of those who use them or are impacted by their actions, and regarding their own security against malicious attacks and wrongful usage.

As we have previously discussed, trust is necessary if these systems are to be widely deployed in our society. Hence, they need to demonstrate *sustained competence* over time, *consistency* in their behaviour so it is reproducible—i.e., it can be replicated providing the same input data and the same situation or context—and *transparency* in their operation, so they are understood by non-experts. We know that Artificial Intelligence algorithms—like most software—are not fool-proof. Thus, another relevant concept is that of *robustness,* which refers to the property that ensures that the behaviour of the system when it is tested on unseen data will be like its behaviour during the training phase, i.e., when exposed to training data. Ideally, the testing error of the system is as close as possible to its training error. Building robust machine learning algorithms is an active area of research, particularly given the rise of the Adversarial Machine Learning area, whose objective is the development of algorithms that would fool existing AI systems.

Our subject of interest are data-driven machine learning-based decision-making systems for social good. These systems need massive amounts of data to be trained on while being increasingly used in domains with consequential impact in people's lives. Hence, the rights to personal data protection and respect for privacy might be questioned and pushed to their limits. There is ample evidence of the misuse of personal data provided by the users of online services, mobile devices and platforms and the aggregation of large amounts of data form different sources by data brokers, with direct implications in people's privacy. It is important to note that machine learning-based algorithms trained on human behavioural data—e.g., data from our interactions with our smartphones and social media platforms—can infer private attributes—e.g., position on the political spectrum, sexual orientation, marital status, education levels and emotional stability—that have never been explicitly disclosed by the users [35]. This ability is essential to understanding the implications of the use of algorithms to model, or influence, human behaviour at the individual level. I believe that certain attributes and traits (e.g., sexual orientation, religion, etc.) should remain in the private sphere and should never be inferred or used by algorithms unless the person expressly consents. The European GDPR includes relevant and pioneering rights, such as the right to establish and develop relationships with other human beings, the right to technological disconnection and the right to be free of vigilance. However, how to enforce the respect of such rights is a non-trivial manner. Other rights that could be added include the right not to be measured, analysed, profiled, influenced, or subliminally manipulated via algorithms or the right to having meaningful human contact, for example, in services operated exclusively by chatbots.

Finally, and as the philanthropy sector very well knows, humans should always be placed at the core. The potential of algorithmic decision-making for philanthropy will only be realised when philanthropic organizations are able to analyse the data, to study human behaviours and to test policies in the real world. A possible way forward is to build living laboratories—communities of volunteers willing to try new ways of doing things in a natural [36]. These could provide a testbed for designing and evaluating algorithmic policymaking approaches that encode societal values.

I believe that it is only when we respect these principles that we will be able to move forward and achieve a model of democratic governance based on data and artificial intelligence, by and for the people. The path forward must place humans and their societal values at the centre of discussions, as humans are ultimately both the actors and the subjects of the decisions made by algorithmic and human means. By involving people and ensuring that their values are respected, we should be able to realise the immense positive potential of data-driven algorithmic decision-making while minimising the risks and the possible negative unintended consequences. The opportunities for the philanthropy sector are countless. I truly hope that we do not miss this opportunity.

References

1. P. Johnson, Global Philanthropy report: perspectives on the global foundation sector. Hauser Institute for Civil Society, at Harvard University (2018).
2. B. Lepri, N. Oliver, E. Letouzé, A. Pentland, P. Vinck, Fair, Transparent, and accountable algorithmic decision-making processes, in *Philosophy & Technology* **31**(4), 611–627 (2017). https://doi.org/10.1007/s13347-017-0279-x.
3. N. Oliver, in *Women shaping global economic governance* (CEPR Press, 2019), pp. 171–181.

4. E. Letouzé, Leveraging Open Algorithms (OPAL) for the safe, ethical, and scalable use of private sector data in crisis contexts, in *Guide to mobile data analytics in refugee scenarios*, ed. by A. Salah, A. Pentland, B. Lepri, E. Letouzé (Springer, 2019), pp. 453–464. https://doi.org/10.1007/978-3-030-12554-7_23.
5. M. Willson, Algorithms (and the) everyday. Inf. Commun. Soc. **20**(1), 137–150 (2016). https://doi.org/10.1080/1369118X.2016.1200645.
6. S. Fiske, in *Handbook of Social Psychology* ed. by S. T. Fiske, D. T. Gilbert, G. Lindzey, (McGraw-Hill, New York, 1998), pp. 357–411.
7. C. Sunstein, Regulation in an uncertain world, at National Academy of Sciences (Washington, DC, June 2012). https://obamawhitehouse.archives.gov/sites/default/files/omb/inforeg/speeches/regulation-in-an-uncertain-world-06202012.pdf.
8. P. Khanna, *Technocracy in America: rise of the info-state* (CreateSpace, Scotts Valley, 2017).
9. J. Froelich, J. Neumann, N. Oliver, Sensing and predicting the pulse of the city through shared bicycling, in *Proceedings of twenty-first international joint conference on artificial intelligence*, pp. 1420–1426 (2009).
10. A. Bogomolov, B. Lepri, J. Staiano, N. Oliver, F. Pianesi, A. Pentland, Once upon a crime: towards crime prediction from demographics and mobile data, at Proceedings of the 16th international conference on multimodal interaction (Istanbul, Turkey, 2014), https://doi.org/10.1145/2663204.2663254.
11. Y. Torres Fernández, D. Pastor Escuredo, A. Morales Guzmán et al., IEEE global humanitarian technology conference (Seattle, Washington, October 2014).
12. B. Lepri, N. Oliver, E. Letouzé, A. Pentland, P. Vinck, Fair, Transparent, and accountable algorithmic decision-making processes, in *Philosophy & Technology* **31**(4), 611–627 (2017). https://doi.org/10.1007/s13347-017-0279-x.
13. S. Centellegher et al., The mobile territorial lab: a multilayered and dynamic view on parents' daily lives, in *EPJ Data Science* **5**, 3 (2016). https://doi.org/10.1140/epjds/s13688-016-0064-6.
14. E. Letouzé, Leveraging Open Algorithms (OPAL) for the safe, ethical, and scalable use of private sector data in crisis contexts, in *Guide to mobile data analytics in refugee scenarios*, ed. by A. Salah, A. Pentland, B. Lepri, E. Letouzé (Springer, 2019), pp. 453–464. https://doi.org/10.1007/978-3-030-12554-7_23.
15. European Commission, Experts say privately held data available in the European Union should be used better and more. https://ec.europa.eu/digital-single-market/en/news/experts-say-privately-held-data-available-european-union-should-be-used-better-and-more.
16. N. Oliver, in *Women shaping global economic governance* (CEPR Press, 2019), pp. 171–181.
17. B. Lepri, N. Oliver, E. Letouzé, A. Pentland, P. Vinck, Fair, Transparent, and accountable algorithmic decision-making processes, in *Philosophy & Technology* **31**(4), 611–627 (2017). https://doi.org/10.1007/s13347-017-0279-x.
18. S. Barocas, A.D. Selbst, Big data's disparate impact, in *California Law Review* **104**(3), 671–732 (2016). https://doi.org/10.2139/ssrn.2477899.
19. D. Pager, H. Shepherd, The sociology of discrimination: racial discrimination in employment, housing, credit and consumer market, in *Annual Review of Sociology* **34**, 181–209 (2008). https://doi.org/10.1146/annurev.soc.33.040406.131740.
20. J. L. Angwin, J. Larson, S. Mattu, L. Kirchner, Machine bias (ProPublica, 2016). https://www.propublica.org/article/machine-bias-risk-assessments-in-criminal-sentencing.
21. C. O'Neil, *Weapons of math destruction: how big data increases inequality and threatens democracy* (Crown, New York, 2016).
22. Y.N. Harari, *21 lessons for the 21st century* (Penguin Random House, New York, 2018).
23. T. Zarsky, The trouble with algorithmic decisions: an analytic road map to examine efficiency and fairness in automated and opaque decision making, in *Science, Technology, and Human Values* **41**(1), 118–132 (2016). https://doi.org/10.1177/0162243915605575.
24. F. Pasquale, *The black blox society: the secret algorithms that control money and information* (Harvard University Press, 2015).
25. J. Burrell, How the machine "thinks": understanding opacity in machine learning algorithms, in *Big Data and Society* **3**, 1 (2016). https://doi.org/10.1177/2053951715622512.

26. Y. LeCun, Y. Bengio, G. Hinton, Deep learning, in *Nature* **521**, 436–444 (2015). https://doi.org/10.1038/nature14539.

27. S. Lluna, J. Pedreira, *Los nativos digitales no existen. Cómo educar a tus hijos para un mundo digital* (Deusto, Barcelona, 2017).

28. A. Andrae, Total consumer power consumption forecast, at nordic digital business summit (Helsinki, Finland, October 2017).

29. E. Pariser, *The filter bubble: how the personalized web is changing what we read and how we think* (Penguin Books, London, 2012).

30. Credit Suisse, Global wealth report. https://www.credit-suisse.com/about-us/en/reports-research/global-wealth-report.html.

31. J. San Pedro, D. Proserpio, N. Oliver, MobiScore: towards universal credit scoring from mobile phone data, in *User modeling, adaptation and personalization. UMAP 2015, Lecture Notes in Computer Science*, vol 9146, ed. by F. Ricci, K. Bontcheva, O. Conlan, S. Lawless (Springer, 2015), pp. 195–207. https://doi.org/10.1007/978-3-319-20267-9_16.

32. M. Hillebrand, I. Khan, F. Peleja, N. Oliver, MobiSenseUs: Inferring aggregate objective and subjective well-being from mobile data, in Proceedings of the European Conference on Artificial Intelligence (ECAI, 2020), pp. 1818–1825.

33. V. Soto, V. Frias-Martinez, J. Virseda, E. Frias-Martinez, Prediction of socioeconomic levels using cell phone records, in *Proceedings of the International Conference on User Modeling, Adaptation, and Personalization* (UMAP, 2011), pp. 377–388.

34. White Paper on Artificial Intelligence—A European approach to excellence and trust. https://ec.europa.eu/info/sites/info/files/commission-white-paper-artificial-intelligence-feb2020_en.pdf.

35. S. Park, A. Matic, K. Garg, N. Oliver, When simpler data does not imply less information: a study of user profiling scenarios with constrained view of mobile HTTP (S) traffic, in *ACM Transactions on the Web* **12**, 2 (2018). https://doi.org/10.1145/3143402

36. S. Centellegher et al., The mobile territorial lab: a multilayered and dynamic view on parents' daily lives, in *EPJ Data Science* **5**, 3 (2016). https://doi.org/10.1140/epjds/s13688-016-0064-6.

Data for Good: Unlocking Privately-Held Data to the Benefit of the Many

Alberto Alemanno

Introduction

When something wrong happens these days—be it a terrorist attack, the break of a pandemic or even a public policy failure—the relevant, yet new question is: who has the data?

Today, having data means being able to save lives. Thanks to our enhanced societal ability to collect, process and use data, we are able to know exactly where—in the aftermath of an episode—the most vulnerable people are, where the epicentre of the disaster is or how a pandemic, such as Zika, is spreading. Most of this information can typically be inferred from where people check their phones following a disaster or which words they use while communicating about it on social media [1].

Yet it is not public authorities who hold this real-time data, but private entities, such as mobile network operators, credit card companies, and—with even greater detail—tech firms such as Google through its globally-dominant search engine, and, in particular, social media platforms, such as Facebook and Twitter.

This chapter is an adapted version, with permission, of an article originally published in the European Journal of Risk Regulation (Alemanno, A., "Data for Good: Unlocking Privately-Held Data to the Benefit of the Many", European Journal of Risk Regulation 2 (2018), available at http://dx.doi.org/10.2139/ssrn.3194040).

Prof. A. Alemanno (✉)
Jean Monnet Professor of EU Law, HEC Paris, France
Member and rapporteur of the European Commission Expert Group on business to government data sharing (2019–2020)
e-mail: alemanno@hec.fr

© The Author(s), under exclusive license to Springer Nature Switzerland AG 2021
M. Lapucci and C. Cattuto (eds.), *Data Science for Social Good*,
SpringerBriefs in Complexity, https://doi.org/10.1007/978-3-030-78985-5_6

Thanks to their many features—including geo-localization and eye-tracking—and our acritical generosity in giving away so much personal information, these companies know more about us than our partners and closest friends [2]. The growing quantity of data harvested by companies through search engines, social networking sites, photo sharing sites, messengers, and apps more generally are the result of the interaction of commercial interests, interface designs, algorithmic processes and users' indication of preferences, actions or attitudes [3].

By supplementing scant public statistics and informing interventions, including in emergencies, big data functions as a crucial 'sense-making resources in the digital era' [4]. In particular, by disclosing these data and mashing them up with various data sets, tech companies may enable public authorities to improve situational awareness and response, thus prioritising interventions (e.g. go where people with asthma are or identify where people from areas hard-hit by a disease are moving to), thus saving money and lives [5]. Think about the kinds of 'safety-check' data from social media applications [6] or mobility patterns stemming from ubiquitous geo-localization services. No doubt this information can supplement scarce public statistics and inform interventions, in particular in emergencies.

But there's more.

While real-time data mostly benefit disaster-zones management—when situations change quickly and life-or-death decisions are typically taken under time-pressure—they may also help a speedy and efficient handling of non-emergency situations. In other words, while humanitarian use of private date epitomises their inherent, yet, locked potential in improving society, their positive societal impact is significantly greater. Indeed, also in less salient areas such as urban planning and health, commercial data present a strong live-improving potential. Thus, for instance, by identifying common predictors for liver failures, you can save lives. By assessing certain trigger messages to teen suicide, public authorities can better assess what preventive approach might (or might not) work. Or by monitoring people's use of public and private transports through geo-tagging, you can generate super-rich data capable of determining how a city operates as well as where and how—for instance—traffic could be improved. These illustrations suggest that, whilst many point out at the risks of data in abstract terms, it is the purpose pursued by data-driven and data-dependent technologies that should draw our attention.

Companies' data can tell us not only whether a given policy intervention works or doesn't work but also how it could be fixed. Thus, policymakers could learn if citizens consume less of an unhealthy product as a result of the implementation of a given policy, be it a soda tax, a health warning or a sale restriction of that product. In other words, access to social media datasets can improve not only the design of new public policies, but also their real-world impact. Therefore data is and must urgently become the hallmark of risk regulation. However, one can't omit to consider that the data we intend to unlock might not cover the so-called 'data invisibles', those people who generally due to their socio-economics, are not counted or tracked within the formal or digital economy [7]. These individuals are disproportionately migrants, women, children, rural and slum dwellers frequently marginalized within their own societies and as such they are not a data point [8].

The Realities of Private Data-Sharing

However, besides a few isolated and self-proclaimed 'data philanthropy' initiatives and other corporate data-sharing collaborations [9], data-rich companies have historically shown resistance to share this data under non-commercial terms. Therefore, despite the undeniable life-changing potential, private data remains the prerogative of a few big corporations who jealously guard it.

While open data is becoming more common in government, academic and institutional datasets [10], this kind of data availability has not yet been taken up by corporations who struggle in embedding these values in their business operations. Thus, for instance, accessibility to private data is not even granted to external actors in, for instance, mandatory data audits, and no public authority is fully aware of the size and nature of the data assets gathered by companies operating in its jurisdiction. More generally, while companies understand the commercial potential of their data, they lack a comparable awareness about the public utility of those data. Social purpose is not how data is wired into their business models and corporate culture. At least not yet, as a rapidly-growing number of companies is increasingly becoming aware of the public—not only the commercial—value of their data assets.

The ensuing phenomenon is giving rise to an alarming data asymmetry in society, which the emerging initiatives of voluntary corporate data sharing do not seem adequate to address and overcome. Such a data asymmetry feeds into consumers' lack of agency, and further strengthens the market dominance of a very few tech corporations, such as Google and its parent company Alphabet Inc., Facebook and subsidiary platforms such as Instagram, Messanger and Whatsapp, and Oculus VR, as well as also increasingly popular apps such as Snapchat.

To justify their restrictive stance, companies typically invoke the need to preserve their competitiveness in the market and to fully protect the privacy of personal information. Moreover, further regulatory, reputational, fiduciary, allocation of risk and other obstacles are typically invoked as preventing companies from sharing their data. As a result, as sharply summed up in the literature:

'Despite the growing acknowledgement of the benefits, we are far from having a viable and sustainable model for private sector data sharing. This is due to a number of challenges—most of which revolve around personal data privacy [11], and corporate market competitiveness [12].'

While several legal and other legitimate obstacles exist to the release of these data, even in emergency situations, those are far from being insurmountable [13]. There are indeed methods of balancing both privacy and competitive risks with data-sharing for public good. These include aggregating data or sharing insights from datasets rather than the raw data. Yet this—as any data-sharing exercise—entails significant transaction costs (e.g. preparing the data, de-risking them, etc) on the supply side. More critically, a broader and deeper cultural shift is needed not only within the corporate sector, but also the public one, so as to pave the way to the emergence of a dedicated ecosystem capable of fostering data collaborations.

The Responsible Data Movement

It is against this backdrop that a growing number of organisations—be they international development, humanitarian, philanthropic as well as other civil society organizations and public authorities—have called (and continue to call) for private digital data to be shared for the public good. They want them to be treated as a 'public goods' because of their inherent value in informing interventions, and not only in emergencies.

The term of data commons was popularised by the *United Nations Global Pulse* initiative, a flagship project promoting the use of big data for sustainable development and humanitarian action [14]. In Davos, during the 2011 World Economic Forum, Kirkpatrick—speaking on behalf of *UN Global Pulse*—complained that '[…] while there is more and more of this [big] data produced every day, it is only available to the private sector, and it is only being used to boost revenues'.

In response to this claim, a 'responsible data' movement has emerged to lay down guidelines and frameworks that will establish a set of ethical and legal principles for data sharing [15]. Yet, in the absence of a common multi-stakeholder platform for data governance, this movement lacks institutionalisation and emerges as a result largely fragmented. Several initiatives, like the Signal Code [16] and the International Red Cross' Handbook on Data Protection in Humanitarian Action [17], are developing. Often based on these frameworks, actors from various sectors—such as social media companies and civil society actors—exchange information to create new public value through so called 'data collaboratives' [18], typically in the form of public-private partnership. While their experimental nature must be praised, these initiatives are—at best—limited to specific sectors, such as health, disaster response, education, poverty alleviation, and—at worst—may hide a commercially-driven attempt at gaining a research insight, access to expertise, entering a new market or merely gaining positive visibility, often visa-vis the companies' employees [19]. As such they remain generally under-used as solutions for large, complex public problems. As a result, no generalizable approach to data sharing has yet emerged [20].

How to then move away from one-off, sectoral projects, toward a scalable, broader, genuine approach to private data-sharing? In other words, how to best "institutionalize" data-sharing for public good within the private sector and in collaboration with public sector and philanthropic actors?

This is the mission pursued by the OECD who has been teaming up with the MasterCard Center for Inclusive Growth to identify and formalize data-sharing methods [21]. After several social media corporations and organizations—such as Facebook, the International Committee of the Red Cross, the United Nations Global Pulse and NYU GovLab —have been exploring various frameworks to private-sector collaborations, the OECD and MasterCard have been analysing dozens of data-sharing initiatives within corporations so as to bring them to the next level. In particular, the OECD has been working on the codification of a set of principles and assessment frameworks aimed at the development of a methodology for voluntary, private-sector data sharing. While this effort must be praised insofar as it goes beyond the sector-by-sector approach, it falls short of developing an accountable governance process capable of applying these principles. Data use and sharing are set to remain highly contingent, contextual and incremental and—more critically—to be entirely subject to data controllers themselves.

The most recent, promising effort at defining what it might become the first public policy for data sharing is led by the European Commission DG Connect that has convened an expert group devoted to business to government data sharing in 2019 [22].

The Cost of Not Sharing Private Data

Moreover, the OECD approach focuses on sharing of insights from corporate data, rather than transfer of raw data to third-party researchers or organizations. While sharing of insights has emerged as a commonly used market model, one may wonder under what circumstances companies will actually accept to do so, given the high-cost involved. More critically, by making data sharing conditional upon a utilitarian cost-benefit analysis ("societal benefits should clearly outweigh other risks") [23], it is unlikely that the proposed OECD approach may alone instil an institutional culture of sharing in the business community.

Shouldn't we ask more often what's the risk of not sharing?

Yet, after more than a decade of the rhetoric of data as public good floating in public and corporate discourse, the idea of systematically assessing the balance between risks and rewards of sharing that data remain the exception—not the norm—in corporate operations. Only few companies—and not necessarily the large ones—have a data management and data governance structures capable of identifying and fostering greater utility for their data or meta-data. More critically only a few of them feel the responsibility of doing so, in co-operation with public authorities.

As a result, despite their potentially life-saving nature, these collaborations are entirely left to the good-will of the private actors involved. In other words, the guardians guard the guardians.

How to Unlock Private Data for Good?

The urgent question therefore today is how to move from this emerging, sector-to-sector, voluntary approach to a more universal, sustainable and accountable data sharing model (or models). To do so, one has to examine whether and how the argument for 'data as a public good' fits with the corporate reality of big data.

It has been suggested that to unlock the potential of commercial data for the public good, one has to move away from the concept of data as something to be owned or controlled. Conferring ownership,—or, as we do in Europe, giving control to third parties over personal information [24]—seems indeed to inherently inhibit societal beneficial use. It is indeed the language of ownership that makes companies blind to the many opportunities where the data they collect and hold create public value.

Yet, even if we were to posit that the data belong to the users and not the company controlling their data, how do we deal with requests for access to the compilation or aggregation of the data of many thousands or millions of users by public authorities claiming they need these data in order to save lives (or attain other legitimate public policy goals)?

A more sensible, emerging approach is that of stewardship, which is intended to convey a fiduciary level of responsibility toward the data. Under such an approach, companies would suddenly reflect how their data can benefit them beyond their businesses' bottom lines, thus embracing a culture of data sharing currently missing. Yet even the stewardship approach does not seem to come to terms to the question of why private sector-data should suddenly be released regardless of an immediate return.

Then what can be done to gain access and use data collected by third parties? Property law is just one of the legal regimes that control right and responsibilities in relation to personal data. Many others exist, such as tort law, contracts, as well as regulatory law, including competition law rules [25]. This suggests that the companies' stubborn refuse to share personal data may lead some of these legal regimes to be triggered soon. Thus, for instance, other companies—be they new entrants on the market, social enterprises or even public authorities—could try to invoke the so-called essential facility doctrine [26]. Despite being defined rather narrowly by the European Court of Justice, some authors are beginning to argue that big data could be regarded—on a case-by-case analysis—as an essential facility, at least in some specific sectors, and as such could be object to mandatory disclosure [27]. The scalability of this model of ad hoc data-sharing would however be inherently limited. A more promising avenue to unlock the power of private-data seems today to be offered by the emergence of a new regulatory regime, notably the EU General Data Protection Regulation (GDPR) and by the supervisory authorities which will oversee its implementation [28].

While its governing principle is that of purpose limitation, which confines the use of data by the data controller to one specific purpose, this Regulation offers some flexibility in given circumstances. It expressly allows a derogation to the principle in case of "further processing for archiving purposes *in the public interest*, scientific or historical research purposes or statistical purposes" [29]. If interpreted favourably and applied proportionally, this provision may be used so as to strike a balance between the interests of data controllers—who could overcome the personal data privacy concerns—and that of the data subjects, whose data could be used in the public interest, to the benefit of the many.

Although the debate over data as public good has not been able to bring about a generalizable model of data sharing, the new European data protection may offer a promising entry point capable of breaking one of the major sources of corporate resistance against data-sharing: personal data privacy and its inherent risk allocation. By making data use accountable, GDPR enables companies (i.e. data controllers) to go beyond the principle of purpose limitation. As such it may grant a supervised yet general power to use personal data beyond the original purpose for which they are collected when it is for the public good. This could potentially be a game-changer in data-sharing practices for the public good. Yet this is not to suggest that fixing the supply side alone by data holders will magically turn data-sharing into a reality.

Also the demand-side by institutions—be they public authorities, civil society organisations, statistical agencies—require some deep structural and methodological work to take full advantage of released data. One indeed can't assume that recipients are ready to use those data. Today, any data-sharing activity requires significant and time-consuming effort and investment of resources for both data holders on the supply side, and institutions that represent the demand.

Hence, the need that both sides of the equation understand one another, align expectations, and that the respective benefits be understood and publicly disclosed. Indeed, both societal and business benefits must be declared and assessed.

Unfortunately, for the time being, most of the efforts focus on getting the supply-side ready—by developing data-share models, de-risking the data, preparing the skill-set for the personnel [30]—for data-sharing, but omit the need to sensitize and prepare the demand side.

Today, establishing and sustaining these new collaborative and accountable approaches requires significant and time-consuming efforts and investment of resources for both data holders on the supply side, and institutions that represent the demand. Yet as more and more data-rich companies are set to realize the utility of their data and understand how the demand side may be using them, this might help to create an enabling environment for data-sharing. And it remains doubtful whether mandating release, on the one hand, or pricing data-sharing, on the other, would per se be capable of attaining such a goal. There is a clear case for continued experimentation.

Conclusions

It is almost a truism to argue that data holds a great promise of transformative resources for social good, by helping to address a complex range of societal issues. Yet as data are inaccessible, especially when it comes to those produced on commercial platforms, it is high time to unlock them for the social good.

The public debate has thus far opposed those who perceive data as commodities and those who believe it is the object of fundamental rights. As demonstrated by this article, the emergent discussion about the use of private data for the public good provides a welcome opportunity to enrich—by making it more complex—such a debate. Time has indeed come to embark on a less polarised conversation when it comes to the governance of data in our societies. To do so it is essential to decouple the legal and ethical aspects of data-sharing and identify new approaches towards it. While the welfare-enhancing properties of data sharing make such a practice a moral necessity, we need technical and legal frameworks capable of translating such a growing moral imperative into workable and legally sound solutions.

At the time in which the conversation about tech companies tends to be negative—as signified by the abuses unveiled by scandals of global-scale tax avoidance or data-harvesting such as *Cambridge Analytica* which were initiated by some 'data philanthropy'—the enduring refusal of the data-rich private sector, notably social media companies, to release part of their data under certain circumstances may trigger a further backlash against them.

Indeed, should the potential of real-time private data to save lives become public-knowledge, the reputation of social media and other data-rich companies could be further tarnished.

How to explain to citizens across the world that their own data—which has been aggressively harvested over time—can't be used, and not even in emergency situations?

Responding to this question entails a fascinating research journey for anyone interested in how the promises of big data could deliver for society as a whole. In the absence of a plausible solution, the number of societal problems that won't be solved unless firms like Facebook, Google and Apple start coughing up more data-based evidence will increase exponentially, as well as societal rejection of their underlying business models.

While embedding data-sharing values in the corporate reality is an opportunity for the few tomorrow, it is already a life-or-death matter for the many, today.

References

1. Today personal data is widely collected by a panoply of largely invisible parties and generally used without the knowledge or consent of exploited 'data subjects.' From the tracking cookies which track our movements on the Web to the flows of data generated by FitBits and other devices, personal data about both our on- and offline activities is harvested, bundled, and monetized daily.
2. You might have heard the story of how a supermarket, Target, worked out a teenage girl was pregnant before her father - and most probably herself - did. Target's consumer tracking system identified 25 products that when purchased together indicate a woman is likely to be expecting a baby. The value of this information was that Target could send coupons to the pregnant woman at an expensive and habit-forming period of her life. That is targeted marketing. To know more, C. Duhigg, How Companies Learn Your Secrets, in The New York Times Magazine (2012), and A. Alemanno, Lobbying for Change: Find Your Voice to Create a Better Society (Iconbooks, London, 2017), pp. 57–59.
3. A. Richterich, *The Big Data Agenda: Data Ethics and Critical Data Studies* (University of Westminster Press, London, 2018).
4. M. Andrejevic, Big Data. Big Questions: The Big Data Divide, in International Journal of Communication **8**, 1675 (2014).
5. For a detailed analysis, see M. Hilbert, Big Data for Development: A Review of Promises and Challenges, in Development Policy Review 34, 1 (2016), p. 135–174, https://doi.org/10.1111/dpr.12142.
6. Facebook declares to share aggregated data about people checking-in safe, using location data of people who have accepted to do so.

7. R. Shuman, F. Mita Paramita, Why your view of the world is riddled with holes (World Economic Forum, 2016), https://www.weforum.org/agenda/2016/01/data-invisibles-ignore-at-our-peril/.
8. India for instance has tens of millions of undocumented immigrants, with 10 million from Bangladesh alone. See Ref. [7]. For an initial collection and taxonomy of corporate data sharing, see GovLab, Data Collaboratives: Creating public value by exchanging data, http://www.dat acollaboratives.org.
9. For an initial collection and taxonomy of corporate data sharing, see GovLab, Data Collaboratives: Creating public value by exchanging data, http://datacollaboratives.org/. As documented by the work of the Open Government Partnership, resistance is also notable in these domains.
10. As documented by the work of the Open Government Partnership, resistance is also notable in these domains.
11. While companies fret to highlight the many privacy and security risks stemming from disclosing personally or demographically identifiable information for the social good, they have shown less concern and care when sharing the very same information for commercial purposes. See, e.g., the investigative work conducted by Carole Cadwalladr, The great British Brexit robbery: how our democracy was hijacked, in *The Guardian* (2017).
12. Pawelke, A. Rima Tatevossian, Data Philanthropy. Where are we Now? (United Nations Global Pulse, 2013), https://www.unglobalpulse.org/2013/05/data-philanthropy-where-are-we-now/.
13. More challenging instead is the issue of data bias, as epitomised by the so called of 'data invisibles', i.e. individuals, generally from vulnerable communities, who are unrepresented in private or public datasets. In other words, the generalisability stemming from extrapolating general observations from such data should be questions.
14. Rima Tatevossian, Data Philanthropy. Public & Private Sector Data Sharing for Global Resilience (United Nations Global Pulse, 2011), https://www.unglobalpulse.org/2011/09/data-philanthropy-public-private-sector-data-sharing-for-global-resilience/. Initially a 'Data for Good' movement has been encouraging using data in meaningful ways to solve humanitarian issues around poverty, health, human rights, education and the environment and now is in the process of being mainstreamed by the OECD.
15. Initially a 'Data for Good' movement has been encouraging using data in meaningful ways to solve humanitarian issues around poverty, health, human rights, education and the environment and now is in the process of being mainstreamed by the OECD.
16. This seeks to apply human rights principles to data during times of emergencies. See https://signalcode.org/.
17. A Handbook was published as part of the Brussels Privacy Hub and ICRC's Data Protection in Humanitarian Action project. It is aimed at the staff of humanitarian organizations involved in processing personal data as part of humanitarian operations, particularly those in charge of advising on and applying data protection standards.
18. UN Global Pulse, GSMA, State of Mobile Data for Social Good, (Report Preview) and the work led by NYU GovLab under this label that has collected more than 100 examples of public value extraction from privately-held data collected. To know more, check http://datacollabor atives.org/.
19. The literature focusing on data-sharing initiatives focusing on Call Details Records (CDR) demonstrates that data sharing occurs where the firm perceives an overall benefit from sharing them, in terms of both business advantage and social impact (which may turn into a business advantage in the longer term). See, e.g. L. Taylor, The Ethics of Big Data as a public good: which public? Whose good?, in *Philosophical Transaction of The Royal Society Publishing* **374** (2016).
20. Noteworthy however is the European Commission's Guidance on sharing private sector data in the European data economy (SWD2018 125 final), which aims to provide a toolbox for companies on the legal, business and technical aspects of data sharing in particular with respect to machine-generated data, notably for public interest purposes.
21. The OECD and Mastercard convened a high-level expert group in data for good in fall 2017 to which I have been a member.

22. Following an open selection process, the Commission has appointed 23 experts to a new Expert Group on Business-to-Government Data Sharing. The group is composed of independent experts with a high level of expertise in the field, covering a broad range of areas of interest and sectors. I am a member of this group.
23. Draft methodology for a Set of Principles/Methodologies for Private Sector Data, OECD, 2018 (on file with the author).
24. Under EU law, this third-party control of data is limited to the rights of the "data subject" which are related to the fundamental right to data protection/privacy.
25. This is insofar as Big data can be seen as a source of market power and, consequently, as a possible field of abuse for undertakings in dominant position. To know more, A. D. Thierer, The Perils of Classifying Social Media Platforms as Public Utilities, in *CommLaw Conspectus—Journal of Communications Law and Policy* **21**, 2 (2013), http://dx.doi.org/10.2139/ssrn.202 5674.
26. Initially developed by the US Supreme Court (United States v. Terminal Railroad Ass'n of Saint Louis, 224 U.S. 383, 1912), this doctrine was applied for the first time in 1993 by the European Commission in a case related to harbor infrastructures before being extended to matters related to nonmaterial facilities.
27. Contra, e.g. G. Colangelo, M. Maggiolino, Big data as misleading facilities. European Competition Journal **13**(2–3), 249–281 (2017). https://doi.org/10.2139/ssrn.2978465.
28. General Data Protection Regulation (GDPR) (EU) 2016/679.
29. Article 5.1 (b) of GDPR.
30. One of the most promising, highly-specific initiative is the one recently undertaken by GovLab with the Data Stewardship Portal available at http://thegovlab.org/tag/data-stewards/.

Building a Funding Data Ecosystem: Grantmaking in the UK

Rachel Rank

Introduction

Grantmaking in the UK takes place in a complex ecosystem of funders and grantees, against the backdrop of the wider social sector. Grant funding is a crucial source of funding for the social sector in the UK, often providing stability and flexibility that is not available from other sources such as fees and contracts or donations from the public.

According to the UK Civil Society Almanac 2019 [1], voluntary organisations spent £7 billion making grants in 2016/17. This doesn't include the grantmaking of government bodies (around £4.6 billion received by voluntary organisations) or National Lottery funders (around £1.6 billion distributed to good causes [2]).

Grantmakers include a wide variety of organisations, including family foundations, corporate philanthropy and government grantmaking. An important part of UK grantmaking is the role played by the 12 National Lottery distribution bodies. These organisations, such as the National Lottery Community Fund or Arts Council England, distribute a proportion of ticket sales from the National Lottery. All levels of government make grants, from small parish or town councils to large central government departments.

Social sector grant recipients can also take many forms. Many grantmakers will concentrate on grants to registered charities, but other forms of non-profit organisations can also be included, such as mutuals or community interest companies. Even public sector bodies such as schools, health services or museums can receive grant funding from a variety of sources.

R. Rank (✉)
Former Chief Executive Officer, 360Giving, London, UK
e-mail: info@threesixtygiving.org

The Information Gap

At the time of writing, it's not possible to find a complete dataset on all charitable grants provided in the UK. This means the huge financial flows—funds from grant-makers to grantees—are opaque. This vastly increases costs for all actors in the sector. Collaboration between grantmakers is made harder, due diligence is done with limited information and grant applicants face significant information barriers to find out who might fund them.

This information gap impacts on all areas of grantmaking in the UK. Central government allocates over £4.6 billion in grants to the voluntary sector every year [3]. Comprehensive grants data would allow us to see which organisations this funding reaches and how it complements the grants made by National Lottery funders, local authorities and the £7 billion of grants made by charitable foundations [4].

The UK's lack of grants data also impacts on emerging trends in public and political discourse. Food banks have become an important part of the UK's response to food poverty over the last decade. But there is no comprehensive data on where they receive funding from, and how this funding has changed.

This also impacts on grantmakers' own development. A grantmaker working in a defined geographical area, for example Wales, would want to develop its strategy based on data that gives a picture of who else is funding in Wales, what they have funded, what the contribution of UK or EU government grants has been and how this funding has changed. Without consistent data from these sources, funders are planning for the future based on their own assumptions about the funding landscape, rather than what is actually happening.

Organisations seeking grants also lose out. A fundraiser working for a small charity needs information to help them decide which funder is best to approach for funding. They need to know what the funders have previously funded, what their strategy and acceptance criteria are. Without this information, fundraisers waste time and resources on inappropriate applications.

Ultimately, these information gaps impact negatively on grantees and, most importantly, their beneficiaries—the people and organisations that grantmakers want to support.

The 360Giving Initiative

360Giving was created to fill this information gap. It was founded in 2014 by UK philanthropist Fran Perrin, who found that much less data was available to inform grantmaking compared to data for making corporate investments. She felt she was "giving in the dark". Fran explained her reasoning in a podcast recorded in July 2018:

> It's not about transparency at all. It's self interest … I want to know what people are doing so I can learn from them, I can collaborate with them and I can get better at my job. It's about making informed choices.

You wouldn't try to do financial investments without any information, with no FTSE 100, with no Reuters. Philanthropy can feel a bit like giving in the dark and I just want us to have better information so we can make the best decisions [5].

360Giving consists of two parts. It is a registered charity that supports grantmaking organisations to publish data about who they fund, and supports those organisations and others to use the data to inform their work. It is also the steward of a data standard—the '360Giving Data Standard'—which provides a common format for sharing data about grants. Both parts of 360Giving—its support work and technical stewardship—are crucial to the initiative's success.

360Giving: The Organisation

360Giving was registered as a company and a charity in 2015, starting with just one member of staff. In 2016 360Giving launched the GrantNav platform [6], which allowed anyone to search the grants data that had been released by the first 20 funders to share their data.

In 2019, 360Giving had a core team of five, and over 100 organisations are sharing funding data using the 360Giving Data Standard.

360Giving works across four main areas, as defined in its 2019–2021 strategy [7]:

1. **Normalise open data sharing**

This involves supporting grantmakers to publish data about the grants they make, both helping grantmakers to publish data for the first time and supporting existing data publishers to continue to update their data. This work is done through providing guidance and tools that help get the data right, and also working directly with grantmakers through a helpdesk.

2. **Improve data quality**

Building on the success of those grantmakers that are already publishing to improve the quality of their data. This could be about helping them to publish additional information about the grants—perhaps including details of the locations of activities. Or it might be about improving the data that they already publish: for example by including a registration number for a recipient organisation so that that record can be linked to extra information about that organisation.

3. **Increase data literacy**

For charitable giving to be more data-informed, it is crucial that grantmaking organisations have the skills and resources to handle and use this data; but the sector has low levels of data literacy. 360Giving is building a community of data users by running events and learning programmes to embed data skills in grantmakers and showcase examples of best practice and innovation.

4.   **Grow data use and shared learning**

The data published by grantmakers only has an impact if it is used. 360Giving supports this by creating tools and resources that make using the data easier, and running events for those who want to use the data. These programmes and resources are tailored to different skill levels, to make sure that the data can be used by those with limited data skills but can also be available to experienced developers and data scientists.

360Giving maintains an active relationship with both data publishers and users in order to help its work. It maintains a list of published data that meets the requirements of the data standard, and it engages with other initiatives that are developing data standards as part of sharing technical approaches and learning.

360Giving: The Data Standard

The 360Giving Data Standard [8] was developed and piloted in 2014 before the organisation was formally registered. A team of open data experts were contracted by the organisation's founder to develop a standard schema. This was then piloted with a test group of grantmakers.

The data standard provides a simple framework for publishing grants data, with flexibility to allow the representation of more complex grantmaking transactions if needed. It is defined as a JSON [9] schema which allows a complex hierarchy of related data objects to be represented; but for most users, this schema can be transformed into a simpler tabular representation that allows the data to be shared through spreadsheets. This tabular format better meets the needs of data publishers and users who do not have advanced data skills. Most data publishers release their data as Excel spreadsheets or Comma Separated Values (CSV) files.

The standard defines a set of 10 fields that must be included for the data to be considered valid. These fields are:

- **Identifier**: A globally unique identifier for each grant.
- **Title**: A title describing the grant.
- **Description**: A more detailed description of the grant activity.
- **Currency**: The currency used for monetary amounts in the grant record.
- **Amount Awarded**: The total amount of the grant.
- **Award Date**: The date the decision to award the grant was made.
- **Recipient Org:Identifier**: A globally unique identifier for the grant recipient.
- **Recipient Org:Name**: The name of the grant recipient.
- **Funding Org:Identifier**: A globally unique identifier for the funder making the grant.
- **Funding Org:Name**: The name of the funder making the grant.

In addition to these required fields, there are some additional (optional) information items that can be provided. These include more information about recipient organisations (for example their address or website url); more detail about grant amounts and transactions (for example the amount that was originally applied for); information about the timing of the grant (for example the time when the grant activity took place); and detailed location information (areas where the grant beneficiaries are located).

Where possible, the data standard builds on existing data standards to produce a system that fits more easily into existing expectations and tools. For example the "Currency" value must be one of the currency codes defined by ISO 4217 and any date fields must use ISO 8601 representation. For identifying organisations, the 360Giving Data Standard recommends the org-id scheme [10], which provides for a globally unique identifier based on existing registration schemes. In practice this means that publishers can use charity or company registration numbers to identify their grant recipients.

The 360Giving Data Standard is supported by a number of tools. The Data Quality Tool [11] is an important part of the data preparation process. This tool checks a file against the standard and provides constructive feedback on where it does not meet the agreed schema, and where standard-compliant data could be enhanced. This tool also allows for data to be easily transformed into different formats, depending on the needs of the users—for example, it can allow an Excel file to be turned into a JSON representation.

While not a formal part of the 360Giving Data Standard, data licensing is an important part of the process for publishing data. Organisations publishing data are encouraged to use an open licence, so that people using the data are not faced with onerous restrictions in accessing it, and others find the data easier to use. This enables innovation—it allows for the development of data analysis and products that reuse the data, even if those products have a commercial purpose. Using an open licence does not mean publishers lose control of their data—they can still include provisions in the licence that require users to attribute and acknowledge the source of the data. 360Giving only includes data with an open licence on the registry of data publishers.

The 360Giving Data Standard has a governance process that ensures that it can change to meet the needs of data publishers and users. This process is led by the Standard Stewardship Committee, which has final decision-making power over changes to the data standard and operates independently of 360Giving the charity, but consults with organisation staff as part of its work. The Committee consists of representative publishers and users of 360Giving data, operating under the agreed Terms of Reference and decision-making process. It considers the impacts that any changes to the Standard would have on users and publishers before deciding whether to proceed. The development of the Standard happens openly, with the minutes of Committee meetings published and changes to the Standard being proposed and discussed through tools such as github.

The Impact of 360Giving

360Giving is a relatively young initiative, so its impact on grantmaking has not yet been fully realised. Some projects and case studies give an indication of the ways in which it has helped civil society. These impacts are discussed here across three themes:

1. Understanding civil society
2. Having data-informed discussions
3. Improving the work of grantmakers and grant seekers.

(1) *Understanding civil society*

Civil society is often misunderstood in the UK. Its size, the contribution it makes to wider society, how many people it employs, how its regulated, how it's funded and by whom—all of these important characteristics can be difficult to find authoritative research on. Open grants data helps people understand and talk about this in a more informed way and makes it easier to use funding data alongside other datasets.

One early example of how the 360Giving dataset could contribute to better under-standing of civil society is a research project conducted by the voluntary sector infras-tructure organisation NCVO in 2015. The project [12], which was funded by Nesta, used an early version of published 360Giving data to find organisations described as "below the radar". Policy discussions often ascribe importance to these organisations that are too small and informal to appear on official registers, such as the register of charities. By their nature they are difficult to find and study, so 360Giving data offers an opportunity to discover some of these organisations.

This research project attempted to find these organisations where they have been funded by a grantmaking foundation. Often this funding might be the only time they appear in published data sources. The work was carried out by taking the list of funded organisations and removing those that did appear on official registers or those that had received a large amount of grant funding. Those that remained were assumed to be "below the radar". The research identified over 33,000 of these organisations, and could then produce information on the kinds of activities they undertook.

Data from the 360Giving corpus can be used to complement and triangulate with existing data sources. The leading statistical resource for the size and scope of UK Civil Society is the NCVO UK Civil Society Almanac (latest edition published in 2019 [13]). This uses data from the annual accounts of registered charities to produce a view of the sector from the perspective of funding recipients. 360Giving data can be used to look at the same transactions from a different angle by focusing on data from the funder side. These two sources reveal different parts of the data picture: data from funding recipients shows how grant funding fits into the rest of their funding, and how they go on to spend the money; while the 360Giving data provides data on individual projects, funding duration and location data which is not available from recipients. 360Giving is particularly keen to include government grants as part of this, due to the important role that funding from government plays for UK civil society.

360Giving data fits well with other data publishing schemes to give a more complete picture of other parts of the funding landscape. Two key complementary schemes are the International Aid Transparency Initiative (IATI) and Open Contracting. IATI [14] is a global initiative to improve the transparency of development and humanitarian resources and their results to address poverty and crises, while Open Contracting [15] helps governments to publish data about the contracts they issue. In both cases the initiatives manage a data standard to help publish relevant data, and all three data standards use the same building blocks (date formats, organisation identifiers, etc.) to help interoperability between them. While 360Giving has been mainly used in a UK context so far, IATI and Open Contracting also demonstrate the potential for international use of data standards, as both are being successfully used across multiple countries.

360Giving data can therefore been seen as an important part of a civil society data ecosystem, both in the UK and beyond. As more publishers share their grants data and with increasing emphasis on data quality, the usefulness of the data for studying civil society grows.

(2)   *Data-informed discussions*

There is an increasing demand within civil society to give communities more of a voice and the need for deeper, closer connections. Better data can help to understand communities, and identify similarities and differences between communities.

In 2018 the Young Foundation published "Patchwork Philanthropy" [16], an examination of the geographical patterns of philanthropic and charitable spending across the UK. The report mapped these patterns against other datasets describing society such as measures of deprivation. The report also focused on the results of the 2016 referendum on whether the UK should leave or remain in the European Union, and examined the differences between areas that voted to leave or remain.

The research used data from 360Giving to provide a picture of philanthropic spending. Data on grants made by funders with a national reach across a specified timeframe was matched with geographical data describing the location of the organisation receiving the grant.

Local Trust, a grantmaking foundation that provides residents with the power to make grants in their own areas, and Oxford Consultants for Social Impact (OCSI) have used 360Giving data as part of the construction of an aggregate index highlighting areas that are "left behind" in their 2019 report "Left behind? Understanding communities on the edge" [17]. Their Community Needs Index brings together data from a range of sources to highlight areas that are lacking in civic assets, are not connected to key services and infrastructure and have low levels of community participation. 360Giving data formed part of the civic assets measures that made up the index, along with data such as community spaces, green space and sport and leisure facilities and average broadband speed.

(3)   *Improving the work of civil society*

It's not just researchers and academics that want information on funding flows. Grant-makers and grant seekers do too. One of 360Giving's key objectives is to provide data that leads to better-informed grantmaking, based on knowledge of where funding gaps are and where resources can be pooled to use them more efficiently.

An important part of this picture is being able to identify recipients that have received funding from more than one funder. This allows funders to work together to ensure they are not duplicating their efforts, or to co-ordinate their funding with others to have a bigger impact. This work is enabled by the 360Giving dataset, particularly through the consistent use of organisation identifiers. Where funders use common identifiers to refer to grant recipients, for example by using a charity or company registration number, it becomes possible to identify those recipients with funding from more than one funder and construct network maps of funding relationships. A 360Giving blogpost shows the process for doing this and visualisations that help to navigate the results [18].

Another area where grantmakers can use data to better understand their own funding is through geographic information. Sharing grants data openly, particu-larly where it includes geographical data, allows for mapping of grants against measures such as levels of deprivation. 360Giving encourages the inclusion of stan-dard geographic identifiers in the published data, through including government-issued area identifiers or through postal codes. This analysis provides great value for grantmakers in understanding where their grants go, but is also straightforward to perform without advanced data analysis techniques, so is accessible to even small grantmakers.

Open grants data can also contribute to a more systematic assessment of grant-making strategy. New Philanthropy Capital (NPC), a charity which supports grant-makers, charities and others to achieve the greatest impact, released a report called "Tackling the homelessness crisis: Why and how you should fund systemically" in 2018 [19]. This report used 360Giving data to look at grant funding for services related to homelessness and used the data to inform recommendations on how homelessness funding could be targeted to be more impactful.

These assessments can also be made using data that covers a geographical area. Analysis of grantmaking in London shows the types and size of funding available to charities based in the city and the sorts of activities that are funded [20]. This allows local infrastructure bodies like London Funders to help their members plan their funding programmes.

Grantmakers can use 360Giving data to look at their work as a whole. An entry to a data visualisation competition held by 360Giving explored how grantmakers fund the core activities of organisations [21]. This work led to a funder exploring core funding in more detail, with recommendations for how funders could better use core funding [22].

These examples show the positive feedback loops from data publishing that allow grantmakers to plan and target their funding more strategically. Grantmakers can feel like they are "giving in the dark"; the data they share can help illuminate the funding landscape.

A Funding Utopia

It should be possible to see where all grant money goes. We should be able to see where in the UK is being underfunded, or overfunded, so that grantmakers—including the government—can better target and complement their grants across the country.

Grantmakers that already publish data about their grants openly, using the 360Giving Data Standard, have begun to see the benefits for themselves and their own operations, as set out in the previous sections of this chapter.

For the full benefits to be felt in society—and for knowledge and power to be better spread through the social sector—all funders must follow suit and publish their grants data openly and consistently to the 360Giving Data Standard, so it can be looked at and compared across regions and sectors.

360Giving's goal is for data sharing to soon become a normal part of the operations of grantmakers, and for use of the data to become embedded in how the social sector operates. By embedding high-quality data within grant funding processes, people on all sides of grant funding transactions—including the ultimate beneficiaries—will be able to make better-informed decisions.

Reaching this 'funding utopia' will not be straightforward. There needs to be technical work to make sure that data is of high quality and that tools are available to support it. Organisations need to invest in skills and resources to properly share and use data about grantmaking and, most importantly, grantmakers, government, civil society and others will need to normalise data sharing and use as part of their culture, to get maximum value from it.

References

1. UK Civil Society Almanac 2019 (NCVO, 2019), https://blogs.ncvo.org.uk/2019/06/19/uk-civil-society-almanac-2019-the-latest-data-on-the-voluntary-sector-and-volunteering/.
2. Where the Money Goes (The National Lottery), https://www.national-lottery.co.uk/life-changing/where-the-money-goes.
3. UK Civil Society Almanac 2019 (NCVO, 2019), https://blogs.ncvo.org.uk/2019/06/19/uk-civil-society-almanac-2019-the-latest-data-on-the-voluntary-sector-and-volunteering/.
4. UK Civil Society Almanac 2019 (NCVO, 2019), https://blogs.ncvo.org.uk/2019/06/19/uk-civil-society-almanac-2019-the-latest-data-on-the-voluntary-sector-and-volunteering/.
5. F. Perrin (founder of the Indigo Trust and 360Giving), Philantropy and Transparency (Giving Thought podcast, 2018). https://givingthought.libsyn.com/fran-perrin-philanthropy-transparency.
6. GrantNav, https://grantnav.threesixtygiving.org/.
7. 360Giving, Our strategy, https://www.threesixtygiving.org/unlocking/0.
8. 360Giving, A standard for sharing grants data, http://standard.threesixtygiving.org/.
9. JavaScript Object Notation (JSON) is a data transfer format which allows for representation of complex data objects, https://en.wikipedia.org/wiki/JSON.
10. org.id-guide, http://org-id.guide/.
11. 360Giving, Data quality tool, https://dataquality.threesixtygiving.org/.

12. Nesta, Mining the grant-makers (2015), https://www.nesta.org.uk/report/mining-the-grant-makers/.
13. 360Giving, Which funders publish grants data openly? http://data.threesixtygiving.org/.
14. International Aid Transparency Initiative, https://iatistandard.org/en/.
15. Open Contracting Partnership, https://www.open-contracting.org/.
16. The Young Foundation, https://youngfoundation.org/publications/patchwork-philanthropy/.
17. OCSI, Left Behind? Understanding Communities on the Edge (OCSI, 2019). https://ocsi.uk/2019/09/05/left-behind-understanding-communities-on-the-edge/.
18. D. Kane, Funding playground—who funds with who in the UK? (360Giving, 2018). http://www.threesixtygiving.org/2018/11/27/funding-playground-who-funds-with-who-in-the-uk/.
19. K. Boswell, R. Tait, Tackling the homelessness crisis: Why and how you should fund systemically (NPC, 2018). https://www.thinknpc.org/resource-hub/tackling-the-homelessness-crisis-why-and-how-you-should-fund-systemically/.
20. London Funders, What can 360Giving data tell us about funding across London? (London Funders, 2019). https://londonfunders.org.uk/our-blog/what-can-360giving-data-tell-us-about-funding-across-london.
21. Suraj Vadgama, Slice & Dice, https://suninthesky.github.io/slice-and-dice/.
22. Esmée Fairbairn Foundation, Insights on core Founding, https://esmeefairbairn.org.uk/latest-news/esmee-insights-on-core-funding/.

A Reflection on the Role of Data for Health: COVID-19 and Beyond

Stefan Germann and Ursula Jasper

Introduction

As previous chapters in this book have thoughtfully and knowledgeably elaborated, data and digitalization have a profound impact on the social, economic and political fabric of contemporary societies. They can be "a powerful tool in making progress on some of humanity's biggest challenges", as Claudia Juech writes in her chapter. At the same time, they pose new risks and problems, with potentially disruptive, destructive consequences.

In this chapter, we aim to shed light on the role of data-based technologies in global public health. COVID-19, the respiratory illness responsible for what is sometimes called the first pandemic of the digital age, has highlighted the many possible uses and applications of data science, digital technology and artificial intelligence (AI) in a public health crisis, including digital epidemiological surveillance, contact tracing, diagnostics, care management and evaluation of interventions. Even before the pandemic, these technologies found their way into health systems and medical science. Because the largest share of these technological developments has been confined to the Global North, we argue that systematic, coherent efforts are needed to tap the benefits of technological progress for people everywhere. Data-based digital technologies and AI are indispensable for improving access to and quality of care especially in resource-poor settings.

Leveraging the potential of digital technologies and AI for everyone requires overcoming technical barriers—from lack of digital infrastructure to missing interoperability of software and data. Efforts must also be shaped carefully and responsibly

Dr. S. Germann (✉)
Chief Executive Officer, Fondation Botnar, Basel, Switzerland
e-mail: sgermann@fondationbotnar.org

Dr. U. Jasper
Governance & Policy Lead, Fondation Botnar, Basel, Switzerland
e-mail: ujasper@fondationbotnar.org

through inclusive, human-centred, participatory processes. Especially in the realm of health, the large-scale collection of data raises difficult ethical, economic, political and legal questions about issues ranging from fairness, non-discrimination and non-stigmatization to benefit-sharing, participation, privacy protection and informational self-determination. To align these concerns, global health data governance principles are desperately needed. But it also needs to be acknowledged that many of these questions are infused with politics, hence they cannot be addressed in a neutral, "clinical" way. In the chapter's final section, we discuss current ethical and governance concerns and suggest how philanthropies could contribute to a public discourse about "a more sophisticated and more considered—and more informed— examination of the pros and cons of datafication … to maximize the public good while limiting harms" (Stefaan Verhulst in his chapter).

The Potential of Leveraging Big Data, Digital Technology and AI in Health

Over the past few years, the transformation brought about by Big Data, digital technology and AI has gained immense worldwide scholarly and media attention. How will these new technologies play out in different policy fields? How will they affect economics and work? And how will they alter the fabric of our societies? While the implications seem diverse, context-dependent and still to a degree unforeseeable, the causes and enabling factors that underlie current developments are much easier to depict.

Broadly speaking, the current technology revolution is facilitated by a convergence of three interrelated developments (Fig. 1), first among them a massive explosion in the amount of data generated and collected on a daily basis. According to estimates by the International Data Corporation, the size of the digital universe grew from 130 exabytes (EB) in 2005 to approximately 40,000 EB in 2020. The World Economic Forum projects that by 2025 "463 exabytes of data will be created each day globally—that's the equivalent of 212,765,957 DVDs per day" [1]. In the health sector, data sources encompass electronic medical records and patient registries, clinical trial data, clinical imaging, administrative and health system management data, genomics data as well as data produced by wearables and social media. Second, as computing power and data storage capacities increase exponentially, computing costs are dropping significantly. For example, today's graphics processing units "can be 40 to 80 times faster than the quickest versions available in 2013" [2]. Third, the set of technologies collectively referred to as AI—including computer vision, natural language processing, virtual assistants and image recognition, robotic process automation and advanced machine learning—are about to become "general purpose technologies": they will revolutionize not just all economic sectors and domains, but societal relations and governance more broadly. On current economic trends, "an estimated 70% of companies might adopt some AI technologies by 2030, up from today's 33%, and

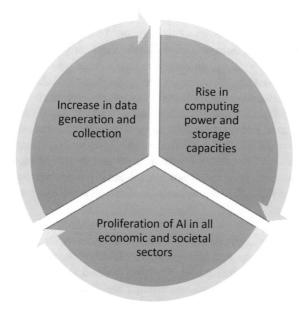

Fig. 1 Convergence of 3 broad trends

about 35% of companies might have fully absorbed AI, compared with 3% today" [3]. The attendant shift in communication patterns and the global growth of digital communication infrastructures are paving the way for what is sometimes called the "Fourth Industrial Revolution" [4].

Data-driven digitalisation, including artificial intelligence and machine learning, is enabling a deep, fundamental transformation of health systems, health services and medical practices in high-income countries and increasingly in low- and middle-income countries (LMICs) too. A recent assessment of the impact of digital technologies on health services suggests that diagnostic, preventative, treatment and rehabilitation options have substantially altered health systems and service provision, especially in terms of structure, culture, professional roles, treatments and outcomes. Internet-connected devices (e.g., the Internet of Things) continue to multiply and improve and, along with Big Data, have become key inputs for innovation, research and development to address current and emerging health challenges.

So far most of this transformation has occurred in high-and-middle-income countries. It now behoves the international community to make the benefits of digital technologies and AI globally available to improve access to and quality of care also in resource-poor settings. According to the World Health Organization (WHO), at least half of the world's people lack access to essential health services today, and despite recent efforts to address the problem, one-third will remain underserved in 2030 unless groundbreaking advances are achieved in the meantime. This outlook is a far cry from the health-related goals articulated in the UN Sustainable Development Agenda in 2015. The global shortage of qualified health workers, which is projected to increase over the coming years, magnifies the problem further, making access to basic quality care even more difficult. As the WHO notes in its Digital Health Strategy, "digital technologies are an essential component and an enabler of sustainable health systems and universal health coverage" [5].

In sum, data-driven digitalization and the application of AI and machine learning are integral to transforming healthcare systems, health services and medical practices and to ensuring progress towards universal access and coverage. These technologies offer unprecedented potential benefits, from clinical decision-making support and remote health worker training to better case management and coordination of care, efficient resource and consultation management and improved access to services, especially for patients living in hard-to-reach areas [6]. Moreover, as COVID-19 has taught us, digital technologies can assist in disease surveillance, outbreak control and contact tracing.

Stronger, more systematic efforts now need to be made—by governments and health authorities, international organizations, funding agencies, philanthropic actors and the private sector, among others—to fully leverage the benefits offered by big data, digital technologies and AI. This requires needs-driven, context-specific, scalable initiatives that are integrated into existing health systems. In global health and development, foundations like the Bill & Melinda Gates Foundation or Bloomberg Philanthropies have recognized the technological potential early on. Merging insights from New Public Management, accounting and epidemiology, these organizations introduced rigorous data-science thinking into their daily work routines and practices to better monitor cost-benefit-ratios of programs and initiatives, to improve resource management and streamline workflows [7]. In the meantime, the digital transformation is not only improving philanthropies' internal management processes, but has also become an objective of many foundation's funding, investment and grant-making work. Today, many foundations and civil-society organizations increasingly promote implementation projects that build upon and integrate digital technology and AI in solutions. Using examples from our work, we will illustrate in the next section how such initiatives, if successfully scaled-up and integrated into health systems, can improve access to and quality of healthcare during COVID-19 and beyond.

Use Cases from the Domain of Global Health: COVID-19 and Beyond

In light of the current global public health crisis of unforeseen proportions, Fondation Botnar decided to commit resources expressly to support international research efforts to advance scientific understanding of and to accelerate the global health response to the outbreak of SARS-CoV-2, the strain of coronavirus responsible for COVID-19, with a particular focus on the role of digital technologies. The public health response to the disease has employed a wide range of approaches and a huge number of specific tools, ranging from population-wide epidemiological surveillance, case identification, contact tracing and public communication to the development and evaluation of diagnostics and interventions.

Of all the digital and AI tools that have been fielded over the course of the pandemic, proximity tracking or digital contact tracing apps have garnered the most attention. At the time of writing, approximately four dozen countries had officially deployed such tools. While the apps differ greatly in terms of underlying technological frameworks, parameters, data sources and "data hunger" as well as degree of privacy protection and voluntariness, they all aim to scale and speed up contact tracing relative to manual methods by automatically matching confirmed cases and their contacts.

Given that automated digital contact tracing has never been used on a such a large scale before, it is hardly surprising that many questions about it remain. Rigorous evaluations of best practices will be essential, also in light of scientists' ever more refined understanding of SARS-CoV-2 and its transmission. There are already strong indications that these apps can be an important tool for interrupting transmission chains and containing the pandemic [8]. For them to work, however, a large proportion of the population must use the respective app. As researchers have pointed out, "adoption is also limited by smartphone ownership, user trust, usability and handset compatibility. Key practical issues remain, such as understanding which contacts are deemed to be close enough for transmission and when exposure time is considered long enough to trigger an alert" [9]. The WHO and many data ethicists add that government authorities and companies that field such apps should adhere to strict ethical and governance principles to ensure non-infringement of core human rights as well as guarantee the protection of key legal values such as proportionality, transparency and accountability [10]. Public health authorities, policy-makers and civil society thus need to agree on a finely calibrated trade-off between the objectives of public health and the protection of individual privacy rights. Otherwise there is a significant risk that trust in digital solutions erodes and parts of the public refrain from using these tools or sharing relevant data. Examples such as the carefully designed, decentralized SwissCovid app indicate that data protection and digital contact tracing can indeed be aligned.

We thus agree with the WHO that "COVID-19-related responses that include data collection efforts should include free, active and meaningful participation of relevant stakeholders, such as experts from the public health sector, civil society organizations, and the most marginalized groups. This participatory approach is not only mandated from an ethics perspective—it will also enhance buy-in, voluntary participation and compliance" [11]. Foundations and philanthropies have an important role to play in this pandemic as enablers and facilitators of such a broad-based, inclusive and transparent public dialogue about how to balance different values and needs to serve the public good, including what constitutes the public good in the first place.

But even before COVID-19 turned the world upside down, Fondation Botnar championed the use of data, AI and digital technology to improve the health and wellbeing of children and young people in fast-growing urban environments around the world and especially in LMICs. Two examples from our work in Tanzania illustrate what such technology-based approaches could look like. Since 2019, together with several implementation partners, we have supported a proof-of-concept initiative called Afya-Tek, which is aimed at integrating digital technologies into a new responsive, people-centered health system in Kibaha, Tanzania. Currently, the Tanzanian health system (like many others, also in high-income countries) is highly fragmented. This not only limits quality of care and access to medical treatments but also results in inefficiencies. Afya-Tek harnesses digital technology to better connect all health system actors—community health workers, health facilities, private drug dispensers and patients—and draws on predictive analytics and biometric identification to improve medical records and decision-making. If successful, this initiative will demonstrate how technology can overcome fragmentation, which is one of health systems' most damaging weaknesses [12].

The other example is the DYNAMIC project, a partnership launched in 2019 with researchers from Unisanté. Every year, 3.3 million children die from an acute febrile episode, and in many countries including Tanzania, increasing use of antibiotics is causing bacterial resistance and ineffectiveness. Studies show that in nine cases out of ten, an antibiotic prescription is not necessary to treat an acute fever in a child. To improve current practices, the project is integrating dynamic clinical algorithms into frontline healthcare to guide and train health workers in the use of antibiotics. Such applications have the potential to improve treatment and to drastically curb unnecessary antibiotic prescriptions. At the public health level, computerized surveillance and collection of epidemiological data in Tanzania will improve early identification of localized epidemics and support both the modification of the patient care algorithm and the implementation of appropriate interventions, such as vaccination campaigns and other preventative measures [13].

Taken together, these examples illustrate how foundations and funding organizations in the field of global public health can promote the use of data science, digitalization and frontier technologies such as AI to transform healthcare.

Difficulties and Challenges: The Need for Data Governance

The benefits of data collection and analysis as a prerequisite for digitalization and frontier technologies and the potential of data-sharing to contribute to significant advances in health are thus widely recognized. Yet key challenges remain. In this section, we shine a light on some of the major ethical and governance issues that must be addressed in order to shape health in the digital age responsibly.

The "3Ms": Missing Data, Misuse of Data and Missed Use of Data [14]

"Missing data" refers to the data gaps that paradoxically persist alongside the overall proliferation of medical and health data. In some sense, not even Big Data is big enough. Missing data are of particularly concern as regards minority groups, people of low economic status who lack access to healthcare and digital technologies and communities where health data are not routinely collected. The resulting biased datasets lead to inadequate computational models and algorithms and, in turn, to flawed health technology applications. Used in medical practice, such applications can lead to defective decisions and thus poor health outcomes. For example, if an algorithm developed to detect skin cancer is only trained with data from light-skinned people then it will fail to arrive at proper diagnoses for persons of other skin colours [15]. Hence, if some populations or groups are underrepresented in datasets, interventions derived from these datasets will not address their health needs appropriately. Studies have shown that missing, biased and unrepresentative data bases can all aggravate existing inequalities in care.

Here, too, the global COVID-19 pandemic has shed new light on an already well-known issue [16]: Schwalbe et al argue, for example, that "hospitals and public health authorities must collect, collate and share disaggregated clinical, demographic and socioeconomic data so that public health professionals can better decode both the clinical risks and understand the risks for various population groups. This will also allow for a response which takes into account social determinants, including population density, mobility, poverty and inequalities when assessing health outcomes" [17]. The missing data problem is made worse by the fact that many countries still lack the necessary infrastructure to collect relevant health data electronically and in a timely manner. Especially in times of crisis, but also under conditions of chronic health worker scarcity, the collection and entry of data into databases can become an additional burden on medical personnel. Unfortunately, failure to collect relevant data is also not always unwitting. Using the example of HIV/ AIDS data, Meg Davis points out that governments sometimes purposefully obfuscate figures or refuse to gather and report data in order to deny the existence of certain groups or problems in their jurisdiction [18].

Second, and relatedly, "missed use of data" describes a situation where data is available but not shared. This can be due to a lack of necessary infrastructure, non-interoperability of systems and protocols or legal-regulatory frameworks and consent mechanisms that could facilitate data sharing. Since there is little international collaboration between the private and public sectors, missed use is particularly likely when data are privately owned, as is the case for an increasing share of health data. The plethora of platforms and sharing partnerships that have emerged within and beyond the health domain in recent years reflect significant investments in data sharing and use. Yet despite the progress in some of the technical and regulatory fields, the world still needs a trusted global framework of fair benefit-sharing that alleviates fears of data exploitation and inequity and facilitates cooperative sharing [19].

"Misuse of data", which refers to unauthorized access to data, can either mean that consent has not been obtained before usage of particular data or that breaches such as fraud and theft have occurred. Because of the sensitive nature of health-related data, regulatory standards as well as technological means of data ownership and privacy protection must be robust. Such measures are also a prerequisite for establishing trust between healthcare institutions and patients because insufficient accountability and transparency will erode people's willingness to share data. While misuse of data is not unique to health, it is particularly important in this domain given the special risk of harm to people whose health data fall into the wrong hands or are not safeguarded from improper sharing.

The Politics of Data

Much of the current debate about Big Data and data science focuses on governance issues and mechanisms to ensure that data collection, use and sharing comply with ethical and legal norms. Ultimately, legally binding global health data governance principles will be needed to align these goals. While the exact design and coverage of such a framework would certainly be subject of extensive discussions, philanthropies, NGOs and civil society actors could support the WHO in advancing such negotiations. One possible outcome could be, for example, a revision/ extension of the International Health Regulations to specifically address the role of data and digital technologies.

But data science and data-based approaches to social and political challenges warrant even more fundamental and principled scrutiny and debate. Challenging questions are not limited to missing data, misuse or missed use of data, but relate to the nature of data and knowledge. We discuss three of those questions here.

First, in debates about the benefits of Big Data and datafication, people tend to implicitly assume that data (about social, economic, political issues, for example) can be collected in a value-free, objective manner—in other words, that the social world can be measured and quantified in a neutral, purely observing way. This epistemology suggests an approach that runs the risk of being "reductionist, mechanistic, atomizing, essentialist, deterministic and parochial, collapsing diverse individuals and complex multidimensional social structures and relationships to abstract data points and universal formulae and laws" [20]. At Fondation Botnar we advocate a relational understanding of people's wellbeing, which holds that material, relational and subjective dimensions together co-constitute wellbeing in a complex, interrelated process. Rather than merely filtering out "data noise" and looking for cause-effect relationships between variables, we want to encourage analyses and initiatives that allow for contextually rich, subjective accounts and "thick descriptions" of the social world [21]. It remains to be seen whether such thick descriptions can be distilled from quantitative data. We hope future research in data science will explore whether and how reductionism and essentialism can be overcome through better data collection, higher processing power and new analytical tools to arrive at more fine-grained analyses that capture context and contingency [22].

Second, we believe that "data advocates" should self-critically acknowledge that the problems they seek to solve, the data they use and the solutions they propose are all fraught with politics. Technology is not a neutral tool, but rather infused with normative decisions about the distribution of resources, power, cultural legacies and political preferences and rights. "Thus, if the field does not openly reflect on the assumptions and values that underlie essential aspects of data science—such as identifying research questions, proposing solutions and defining 'good'—the assumptions and values of dominant groups will tend to win out. Projects that purport to enhance social good without a reflexive engagement with social and political context are likely to reproduce the exact forms of social oppression that many working towards 'social good' seek to dismantle", Green writes [23].

Third, perhaps one of the most radical criticisms of the ongoing digital transformation of whole societies has been levelled by Philip Alston, UN Special Rapporteur on Extreme Poverty and Human Rights. In his 2019 report to the UN General Assembly, he warns that a lack of careful, informed analysis and public deliberation about the costs and benefits of digital transformation puts governments and societies all over the world at risk of "stumbling, zombie-like, into a digital welfare dystopia" [24], where citizens—especially welfare recipients and the powerless—become ever more monitorable by their governments (and by private sector companies, we might add). Alston maintains that "values such as dignity, choice, self-respect, autonomy, self-determination and privacy" [25] are sacrificed for the sake of digital transformation.

Such dramatic assessments underestimate the positive, empowering function of digital tools. Yet they should caution us against uncritical digital euphoria. The digital transformation allures because it is associated with positive attributes such as modernity, progress, resource efficiency and managerial oversight and is often seen as an unpolitical, technocratic solution to large-scale challenges. While some of these attributes might be (or become) real, it is still important to arrive at a more balanced understanding of the pros and cons and of the more fundamental societal shifts that will characterize life in the digital age. This requires a substantial public understanding of data and the digital transformation of societies, as well as more inclusive, participatory decision-making processes to tackle questions surrounding them. Foundations and other civil society organizations should support education programs to foster deeper digital literacy and act as facilitators, conveners and enablers of an informed broad-based and inclusive public dialogue about the politics involved.

Conclusion

It is today a widely shared view that digital technologies and AI, if made available globally, can be powerful tools to address some of humanity's biggest challenges.

By drawing on examples from our own work, we have illustrated that data-driven digitalization and the application of AI and machine learning are integral to transforming healthcare systems, health services, research and medical practices. While the largest share of these technological developments has so far been confined to the Global North, these technologies are indispensable for improving access to and quality of care especially in resource-poor settings.

However, in the realm of digital health, the large-scale collection of data raises difficult ethical, economic, political and legal questions about how to guarantee human rights and core values such as fairness, non-discrimination and non-stigmatization, benefit-sharing, participation, privacy protection and informational self-determination. A global health data governance framework would help to align the different objectives of data collection, sharing, use and protection. But there is also a need for a more critical assessment of the nature of data and the knowledge we derive from Big Data: not only do we need to acknowledge the political nature of data and data science; we must also avoid reducing diverse individuals and their multifaceted social structures and relationships to mere data points.

The ongoing technological revolution is having a profound impact on the social, economic and political fabric of contemporary societies. It will be crucial to shape these developments carefully and responsibly through inclusive, human rights-based, participatory processes. We are convinced that philanthropies and civil society actors have a particularly important role to play in this journey—as enablers and facilitators of a broad and transparent public dialogue about how to shape a fair and equitable digital ecosystem of the future.

References

1. World Economic Forum, How much data is generated each day? (WEForum, 2019). https://www.weforum.org/agenda/2019/04/how-much-data-is-generated-each-day-cf4bddf29f/.
2. ITUTrends, Assessing the economic impact of artificial intelligence (ITU, 2018), p. 1. https://www.itu.int/dms_pub/itu-s/opb/gen/S-GEN-ISSUEPAPER-2018-1-PDF-E.pdf.
3. ITUTrends, Assessing the economic impact of artificial intelligence (ITU, 2018), p. 23. https://www.itu.int/dms_pub/itu-s/opb/gen/S-GEN-ISSUEPAPER-2018-1-PDF-E.pdf.
4. K. Schwab, The fourth industrial revolution. What it means and how to respond (Foreign Affairs, 2015). https://www.foreignaffairs.com/articles/2015-12-12/fourth-industrial-revolution.
5. World Health Organization, Global Strategy on digital health (WHO), https://www.who.int/docs/default-source/documents/gs4dhdaa2a9f352b0445bafbc79ca799dce4d.pdf.
6. US Agency for International Development, Artificial Intelligence in Global Health (USAID), https://www.usaid.gov/sites/default/files/documents/1864/AI-in-Global-Health_webFinal_508.pdf.
7. D. Reubi, Epidemiological accountability: philanthropists, global health and the audit of saving lives. Econ. Soc. **47**(1), 83–110 (2018). https://doi.org/10.1080/03085147.2018.1433359.
8. L. Ferretti et al., Quantifying SARS-CoV-2 transmission suggests epidemic control with digital contact tracing. Science **368**, 6491 (2020). http://dx.doi.org/10.1126/science.abb6936.
9. J. Budd et al., Digital technologies in the public-health response to COVID-19. Nat. Med. **26**, 1183–1192 (2020). https://doi.org/10.1038/s41591-020-1011-4.
10. World Health Organization, Ethical considerations to guide the use of digital proximity tracking technologies for COVID-19 contact tracing (WHO), https://apps.who.int/iris/bitstr

eam/handle/10665/332200/WHO-2019-nCoV-Ethics_Contact_tracing_apps-2020.1-eng.pdf? sequence=1&isAllowed=y; see also F. Lucivero et al., COVID-19 and contact tracing apps: ethical challenges for a social experiment on a global scale. J. Bioethical Inquiry **17**, 835–839 (2020). https://doi.org/10.1007/s11673-020-10016-9.

11. World Health Organization, Ethical considerations to guide the use of digital proximity tracking technologies for COVID-19 contact tracing (WHO), p. 5, https://www.who.int/publications/i/item/WHO-2019-nCoV-Ethics_Contact_tracing_apps-2020.1.

12. Fondation Botnar, Fondation Botnar and partners launch Afya-Tek, https://www.fondationbotnar.org/project/afyatek/.

13. Fondation Botnar, Fondation Botnar announces a new partnership with Unisanté (Fondation Botnar, 2019). https://www.fondationbotnar.org/fondation-botnar-announces-a-new-partnership-with-unisante/.

14. We owe this typology to Amandeep Singh Gill, Director of the International Digital Health and AI Research Collaborative (I-DAIR), Geneva.

15. A. Lashbrook, AI-driven dermatology could leave dark-skinned patients behind (The Atlantic, 2018). https://www.theatlantic.com/health/archive/2018/08/machine-learning-dermatology-skin-color/567619/.

16. E. M. Cahan et al., Putting the data before the algorithm in big data addressing personalized healthcare. NPJ Digital Med. **2**, 78 (2019). https://doi.org/10.1038/s41746-019-0157-2.

17. N. Schwalbe et al., Deaths from covid-19 could be the tip of the iceberg (The BMJ Opinion, 2020). https://blogs.bmj.com/bmj/2020/05/20/deaths-from-covid-19-could-be-the-tip-of-the-iceberg/.

18. S. L. M. Davis, *The Uncounted. Politics of Data in Global Health* (Cambridge University Press, 2020).

19. N. Couldry, U. A. Mejias, *The Costs of Connection. How Data is Colonizing Human Life and Appropriating it for Capitalism* (Stanford University Press, 2019).

20. R. Kitchin, The ethics of smart cities and urban science. *Philosophical Transactions of the Royal Society*, **374**, 4, 2083 (2016). http://dx.doi.org/10.1098/rsta.2016.0115.

21. S. White, Bath Papers in International Development and Wellbeing 43 (2015), Relational wellbeing: A theoretical and operational approach. https://www.econstor.eu/bitstream/10419/128138/1/bpd43.pdf.

22. R. Kitchin, The ethics of smart cities and urban science. *Philosophical Transactions of the Royal Society*, **374**, 4, 2083 (2016). http://dx.doi.org/10.1098/rsta.2016.0115.

23. B. Green, Data Science as Political Action. Grounding Data Science in a Politics of Justice, p. 20 (2020). https://scholar.harvard.edu/files/data_science_as_political_action_v2.pdf.

24. P. Alston for the United Nations General Assembly, Extreme poverty and Human rights (2019). https://undocs.org/A/74/493.

25. P. Alston for the United Nations General Assembly, Extreme poverty and Human rights (2019). https://undocs.org/A/74/493.

Printed in the United States
by Baker & Taylor Publisher Services